AF786956

Advanced Nano-Structured Hard Coatings: Design, Synthesis, and Applications

Advanced Nano-Structured Hard Coatings: Design, Synthesis, and Applications

Guest Editors

Hongbo Ju
Bingyang Ma
Yuxin Wang
Filipe Fernandes

MDPI

Basel • Beijing • Wuhan • Barcelona • Belgrade • Novi Sad • Cluj • Manchester

Guest Editors

Hongbo Ju
School of Materials Science and Engineering
Jiangsu University of Science and Technology
Zhenjiang
China

Bingyang Ma
School of Materials Science and Engineering
Shanghai Dianji University
Shanghai
China

Yuxin Wang
School of Materials Science and Engineering
Jiangsu University of Science and Technology
Zhenjiang
China

Filipe Fernandes
CIDEM, ISEP
Polytechnic of Porto
Porto
Portugal

Editorial Office
MDPI AG
Grosspeteranlage 5
4052 Basel, Switzerland

This is a reprint of the Special Issue, published open access by the journal *Coatings* (ISSN 2079-6412), freely accessible at: https://www.mdpi.com/journal/coatings/special_issues/Ahard_coatings.

For citation purposes, cite each article independently as indicated on the article page online and as indicated below:

Lastname, A.A.; Lastname, B.B. Article Title. *Journal Name* **Year**, *Volume Number*, Page Range.

ISBN 978-3-7258-4797-6 (Hbk)
ISBN 978-3-7258-4798-3 (PDF)
https://doi.org/10.3390/books978-3-7258-4798-3

© 2025 by the authors. Articles in this book are Open Access and distributed under the Creative Commons Attribution (CC BY) license. The book as a whole is distributed by MDPI under the terms and conditions of the Creative Commons Attribution-NonCommercial-NoDerivs (CC BY-NC-ND) license (https://creativecommons.org/licenses/by-nc-nd/4.0/).

Contents

coatings

MDPI

Article

Fabrication and Tribological Properties of Diamond-like Carbon Film with Cr Doping by High-Power Impulse Magnetron Sputtering

Shuai Liu [1], Wenjian Zhuang [1], Jicheng Ding [1,2,*], Yuan Liu [1], Weibo Yu [3], Ying Yang [1], Xingguang Liu [1], Jing Yuan [4] and Jun Zheng [1,*]

1 Key Laboratory of Green Fabrication and Surface Technology of Advanced Metal Materials, Ministry of Education, Anhui University of Technology, Maanshan 243002, China; liushuaiahut@126.com (S.L.); zwj23auot@126.com (W.Z.); liuy202309@nuaa.edu.cn (Y.L.); yangying@ahut.edu.cn (Y.Y.); sdwfcllxg@126.com (X.L.)

2 China International Science and Technology Cooperation Base on Intelligent Equipment Manufacturing in Special Service Environment, Anhui University of Technology, Maanshan 243002, China

3 Changzhou Huachuang Aviation Technology Co., Ltd., No. 539 Xiacheng South Road, Wujin District, Changzhou 213161, China; yuweibo1987@126.com

4 Anhui HERO Electronic Sci & Tec Co., Ltd., No. 129 West Section of Cuihu Fifth Road, Economic and Technological Developent Zone, Tongling 244002, China; hero88001@163.com

* Correspondence: jcdingxinyang@126.com (J.D.); zhengj_ahut@163.com (J.Z.)

Abstract: The present study aims to investigate the advantages of diamond-like carbon (DLC) films in reducing friction and lubrication to address issues such as the low surface hardness, high friction coefficients, and poor wear resistance of titanium alloys. Cr-doped DLC films were deposited by high-power impulse magnetron sputtering (HiPIMS) in an atmosphere of a gas mixture of Ar and C_2H_2. The energy of the deposited particles was controlled by adjusting the target powers, and four sets of film samples with different powers (4 kW, 8 kW, 12 kW, and 16 kW) were fabricated. The results showed that with an increase in target power, the Cr content increased from 3.73 at. % to 22.65 at. %; meanwhile, the microstructure of the film evolved from an amorphous feature to a nanocomposite structure, with carbide embedded in an amorphous carbon matrix. The sp^2-C bond content was also increased in films, suggesting an intensification of the film's graphitization. The hardness of films exhibited a trend of initially increasing and then decreasing, reaching the maximum value at 12 kW. The friction coefficient and wear rate of films showed a reverse trend compared to hardness variation, namely initially decreasing and then increasing. The friction coefficient reached a minimum value of 0.14, and the wear rate was 2.50×10^{-7} $(mm^3)/(N{\cdot}m)$, at 8 kW. The abrasive wear was the primary wear mechanism for films deposited at a higher target power. Therefore, by adjusting the target power parameter, it is possible to control the content of the metal and sp^2/sp^3 bonds in metal-doped DLC films, thereby regulating the mechanical and tribological properties of the films and providing an effective approach for addressing surface issues in titanium alloys.

Keywords: HiPIMS; DLC films; Cr doping; microstructure; tribological property

1. Introduction

Titanium alloy materials, as lightweight materials with high strength and a strong corrosion resistance, have broad application prospects [1,2]. For example, in the aerospace field, titanium alloys are used for manufacturing the structural components of aircraft, engine parts, etc., due to their high strength and low density, which can reduce the aircraft weight and improve fuel efficiency [3–5]. In the automotive field, titanium alloys can be used to manufacture high-performance automotive components, enhancing the performance and safety of vehicles. In the medical device field, titanium alloys are widely used in the production of artificial joints, dental implants, etc., due to their excellent biocompatibility and corrosion resistance, reducing the harmful impact on the human body [6]. However, titanium alloys still face challenges regarding their surface hardness and wear resistance, limiting their use in certain applications. To overcome the shortcomings of the low surface hardness and insufficient wear resistance of titanium alloys [7–9], Physical Vapor Deposition (PVD) technology, with its low temperature and environmentally friendly advantages, can be an effective surface modification method. By depositing PVD films on the surface of titanium alloys without affecting their overall excellent performance, a surface modification treatment can impart high hardness, a low friction coefficient, and excellent wear resistance to titanium alloys, effectively addressing their performance deficiencies [10].

Diamond-like carbon (DLC) films, commonly fabricated by the PVD method, are mainly composed of a metastable structure consisting of sp^3 and sp^2 hybrid carbon bonds, exhibiting good mechanical properties and anti-friction and wear characteristics [11,12]; they are widely applied in the aerospace, hydrogen energy, and biomedical fields [13–17]. Regardless of the type of DLC film, the effective control of process parameters and modifications to the film structure can significantly enhance its anti-friction lubrication properties. This approach effectively addresses the issues of low hardness and severe wear on the surface of titanium alloys. However, pure DLC films have significant internal stress, making them prone to delamination from the substrate. Doping films with metal elements can not only reduce residual stress and improve film/substrate adhesion but also enhance their hardness and frictional properties [18]. Moreover, doping DLC films with metal atoms can create a special nanostructure, endowing the film with the exceptional mechanical and tribological properties that pure DLC films lack. Researchers such as Ding et al. [19] have prepared DLC films with different Nb contents using hybrid magnetron sputtering technology, finding that the addition of metal Nb significantly reduces the internal stress of the film. By adjusting the Nb content, the proportion of sp^2/sp^3 bonds in the film can be altered, thereby regulating the film's mechanical properties. Guo et al. [20] used HiPIMS technology to prepare hydrogen-free Al-DLC films using AlC composite targets at different bias voltages. The results showed that the hardness of the film was significantly varied by changing the sp^3-C bond content through bias voltage adjustments. Santiago et al. [21] found that doping DLC films with 3 at. % Cr made it easier to form a carbon-containing layer on the friction contact surface, leading to a superior high-temperature wear performance during the friction process.

There are various PVD methods for preparing DLC films; among them, direct current magnetron sputtering (DCMS) and arc ion plating (AIP) are the most common PVD sputtering technologies. However, due to its low ionization rate during deposition, DCMS technology may lead to poor film–substrate adhesion [22]. AIP technology tends to produce large particles on a surface, resulting in poor surface qualities and tribological properties. Compared to DC magnetron sputtering and arc ion plating, high-power impulse magnetron sputtering (HiPIMS) technology exhibits higher ionization rates due to the application of very high peak power densities for extremely short periods within a single pulse cycle, significantly enhancing the bombardment and etching effects of deposited ions [23,24]. The energy and density of the participating film-forming ions are higher, which results in a higher film–substrate bonding strength and smoother, denser films with fewer defects and large metal particles [25]. Meanwhile, the energy, flux, performance, and incident angle of the bombarding ions also play a significant role in regulating the film's microstructure and mechanical properties during film growth. Kouznetsov et al. [26] also proposed an enhancement of the target atom ionization rates by increasing the target power density, thus improving the comprehensive performance of deposited films.

In this work, DLC films with different Cr contents, obtained by adjusting the target power, were deposited using HiPIMS technology. The effects of power parameters on the doping content, microstructure, and mechanical and tribological properties of the films were investigated in detail. The relationship between the process parameters and film structure properties was established, achieving a controllable film structure and providing a new method for the surface modification of titanium alloys.

2. Experiment Details

DLC films were deposited using an industrial-scale sputtering system with Q-Plex coating equipment with the scale of 700 mm × 700 mm × 180 mm. The cylindrical Cr target (ϕ 155 mm × 660 mm, 99.99% purity) was powered by an HiPIMS unit. Titanium alloy (TC4) single-sided polished samples (ϕ 32 mm × 3 mm) and single-crystalline Si (100) wafers were used as experimental substrates. To further remove some contamination on the substrate, prior to deposition, the substrates underwent ultrasonic cleaning in acetone and alcohol for 30 min. Initially, a Cr/CrN transition layer was deposited, followed by the deposition of the Cr-DLC layer. The purpose of the transition layer is to enhance the adhesion between the film and the substrate. The DLC layer deposition temperature was 180 °C, with a substrate bias voltage of −100 V, maintaining a deposition pressure of 0.5 Pa and a deposition time of 90 min. The films were prepared under uniform conditions with an Ar/C_2H_2 ratio of 1:2, and the HiPIMS power frequency was fixed at 4000 Hz. The film structure was adjusted by varying the target power from 4 kW to 16 kW. The specific deposition process parameters for the DLC layer and the transition layer are shown in Table 1.

Table 1. Deposition parameters of Cr-DLC films.

Parameters	Cr	CrN	Cr-DLC
Base pressure (Pa)	5.0×10^{-5}	5.0×10^{-5}	5.0×10^{-5}
Working pressure (Pa)	0.5	0.5	0.5
Bias voltage (V)	−50	−50	−100
HiPIMS power (kW)	15	15	4, 8, 12, 16
Pulse length (μs)	100	100	70
Flow rate (sccm)	Ar: 300	Ar/N_2: 200/100	Ar/C_2H_2: 150/300
Deposition time (min)	15	60	90

A field emission scanning electron microscope (FE-SEM, FEI Nano430, Amsterdam, The Netherlands) was used to observe the surface and cross-sectional morphologies of films, measuring thickness and calculating the deposition rate of films. The X-ray energy-dispersive spectrometer (EDS) attached to the FE-SEM was used to measure the elemental content in Cr-DLC films. A surface profilometer was used to detect film surface roughness (Sa), with a measurement range of 500 μm × 500 μm. A Raman spectrometer was used to analyze the film bonding characteristics. For Raman testing, a wavelength of 532 nm was selected, with wavenumbers ranging from 800 to 2200 cm^{-1} and a minimum spot diameter of 1 μm.

The chemical bonding state was determined using X-ray photoelectron spectroscopy (XPS, Escalab 250Xi, Waltham, MA, USA) using mono Al Kα (hv = 1486.6 eV) at 15 kV and 15 mA. Prior to XPS signal acquisition, the surfaces of films were cleaned with Ar^+ for 120 s. The C 1s binding energy of 284.88 ± 0.14 eV was used for the calibration. The hardness (*H*) and elastic modulus (*E*) of films were tested using a nanoindenter. To ensure reliable testing results were obtained, measurements were taken at ten areas on the film surface, and the average value was recorded. The wear performance of films was evaluated using a ball on disk tribometer. The test conditions included a 9Cr18 steel ball with a diameter of 6.35 mm, a load of 5 N, a linear velocity of 30 cm/s, a rotation radius of 5 mm, and a relative humidity of 40%~60%. After the wearing test, the films were observed using a step profiler and an optical microscopy. Finally, the wear rate (W) was calculated based on the Archard equation:

$$W = V/(N \times L)$$

where V is volume loss, N is the normal load, and L is the total sliding distance.

3. Results and Discussion

The composition of Cr-DLC films prepared at different powers is illustrated in Figure 1. It can be observed that, as the target power increases from 4 kW to 16 kW, the Cr content rises from 3.73 at. % to 22.65 at. %, while the C content exhibits the opposite trend, decreasing from 94.28 at. % to 75.44 at. %. As the target power increases, the HiPIMS peak power density also rises, enabling secondary electrons on the target surface to acquire higher energy and speed. This increases the probability of collisions with the target material, thereby enhancing the Cr ionization rate [27]. Throughout the whole deposition process, the deposition state remains stable, with no apparent arcing on the target surface. This is a significant advantage of the HiPIMS power source, effectively suppressing/delaying arcing compared to traditional DC power sources. Therefore, by adjusting the target power, it is possible to control the Cr doping content to range from trace amounts to high levels. Additionally, an O content below 5 at. % was detected in all films; this could be ascribed to the residual air in the vacuum chamber. The relationship between the film thickness and deposition rate of the films with power variations is shown in Figure 2. As the power increases, the film thickness grows from 1.40 μm to 2.06 μm, and the corresponding deposition rate rises from 15.56 nm/min to 22.90 nm/min. This is due to the increased power leading to higher ion sputtering rates on the target material [28], thereby increasing the probability of collisions with C_2H_2 molecules and subsequently improving the deposition rate of the film. Further increases in power do not significantly alter the deposition rate. Although an excessive target power can improve the ionization rate, increasing the negative potential of the target will attract the ionized target ions back to the target material, which is the self-sputtering effect [29], meaning that there is little change in the deposition rate after the power increases beyond 12 kW.

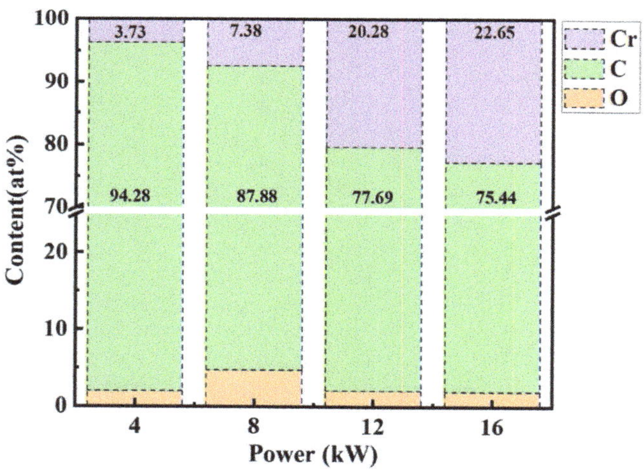

Figure 1. The chemical composition of the films as a function of power.

The surface and cross-sectional morphologies of the films with various target powers were characterized using SEM, as shown in Figures 3 and 4. At 4 kW, the film surface appears smooth and dense, with no significant defects except some particles. As for 8 kW, spherical precipitates are visible on the film surface and these small precipitates are uniformly distributed. At 12 kW, a notable change in morphology occurs, with a transition from cluster formations to fine scale-like structures. At the same time, the film's surface becomes denser, the solid solubility of Cr in the film reaches its limit, and the Cr-rich phase begins to precipitate on the surface. Upon reaching 16 kW, the scale-like structures merge into larger tissues, and the Cr content in the film also reaches the maximum. The corresponding cross-sectional morphologies of films are depicted in Figure 4, showing the films have a uniform growth distribution at all powers and good adhesion with the substrate interface. The transition layer and DLC layer exhibit an obvious layered structure. As seen in Figure 4a–d, a prominent columnar crystal growth is observed in the transition layer, which is adherent with the DLC layers. With the increase in target power, the thickness of DLC layer increases and the structure shows no obvious changes. The three-dimensional surface morphologies of films with different target powers are shown in Figure 5. As the target power increases, the surface roughness (Sa) of the film decreases from 312.0 nm at 4 kW to 48.3 nm at 8 kW, followed by a slight increase, but stabilizes thereafter. This is attributed to the lower ion energy at lower target powers, which affects the Cr atoms that are uniformly doped in films, leading to an uneven film growth and increasing the probability of the formation of large particles [30]. At a relatively high target power, Cr atoms can achieve sufficient energy for diffusion and migration on the surface of the film, resulting in a smoother and denser film surface. However, with further increases in the target powers, the Cr content in films reaches its solid solution limit, and thus the precipitation of the Cr-rich phases increases the surface roughness. When the number of precipitates increases and they aggregate together, the surface roughness is reduced. Consequently, the Sa value decreases from 147.4 nm at 12 kW to 84.9 nm at 16 kW.

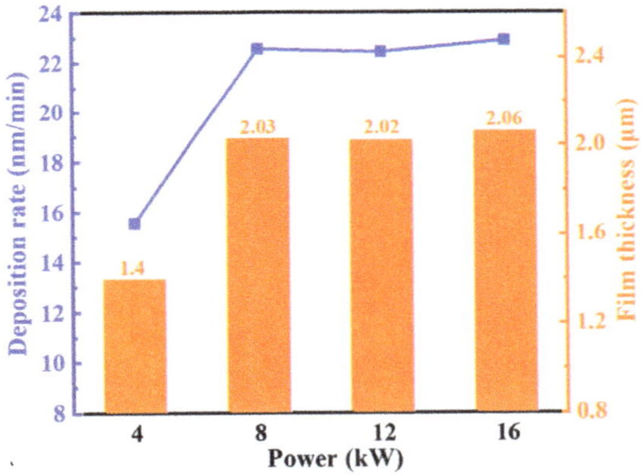

Figure 2. Deposition rate and thickness of the films under different target powers.

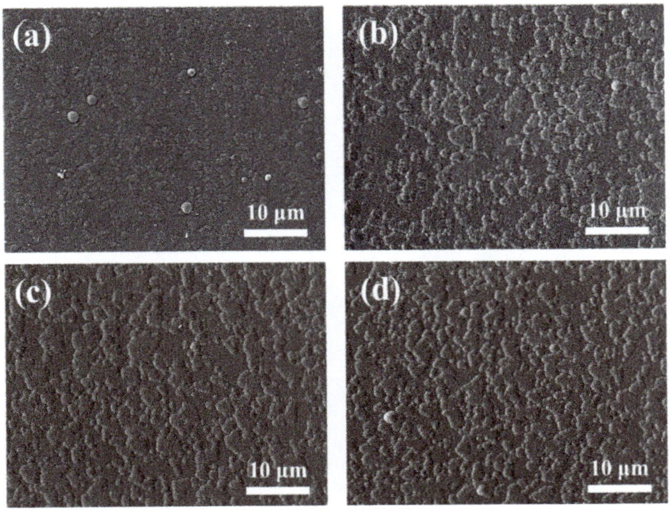

Figure 3. Surface images of films with various target powers: (**a**) 4 kW; (**b**) 8 kW; (**c**) 12 kW; and (**d**) 16 kW.

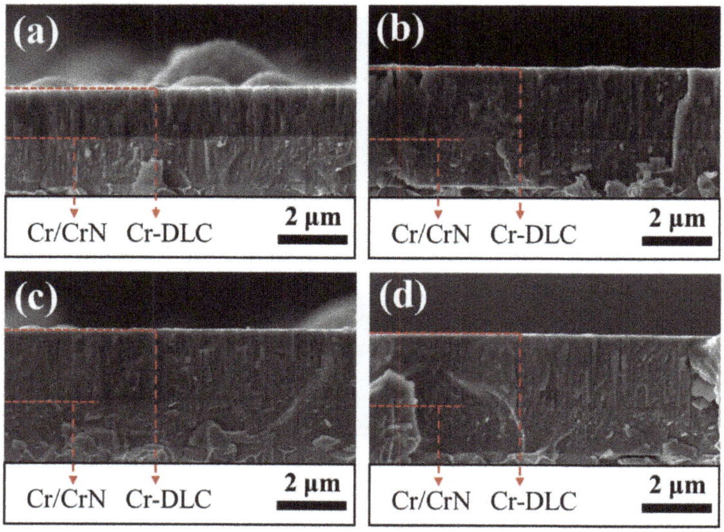

Figure 4. Cross-sectional images of films with various target powers: (**a**) 4 kW; (**b**) 8 kW; (**c**) 12 kW; and (**d**) 16 kW.

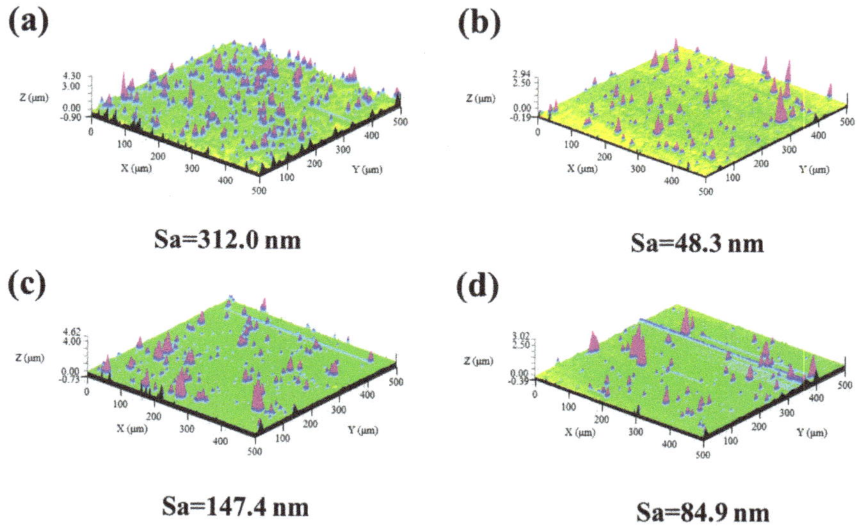

Figure 5. Surface roughness of films with different target powers: (**a**) 4 kW; (**b**) 8 kW; (**c**) 12 kW; and (**d**) 16 kW.

The Raman spectra of films with different target powers are shown in Figure 6. As the power increases, the Raman spectrum signal intensity gradually diminishes, with smaller and weaker peaks. This effect was also observed in previous studies when doping DLC with other metals, such as V and Zr [21,31]. This can be ascribed to the reduction in the light absorption capability of the DLC films when introducing metal dopants. Meanwhile, the intensity of the decrease in the Raman spectra also suggests that the C content decreased in films, which is consistent with the results of Figure 1. The common feature among the four film groups is the appearance of an asymmetric broad peak in the range of 1000–1800 cm^{-1}, with a minor shoulder peak in the low-wave-number region, typical of amorphous DLC film Raman spectra. To further analyze the amorphous carbon bonding structure in the films, Gaussian fitting was employed to fit the Raman spectra into two Gaussian peaks: the D−peak at around 1360 cm^{-1} due to the breathing vibration of sp^2 atoms, which occurs only in rings, and the G−peak at around 1580 cm^{-1} due to the vibrations of sp^2 atoms in both rings and chains, as shown in Figure 6b,c. A high degree of overlap between the fitting curve and the Raman spectrum curve was obtained, indicating accurate and reliable fitting results. With an increase in target power, the Cr doping level rises, leading to a larger D peak area relative to the G peak area. This signifies an increase in the sp^2/sp^3 bond ratio [32]. This is due to Cr doping promoting an increase in the sp^2−C bond content and the degree of graphitization. Figure 6d presents the I_D/I_G ratios with respect to the target power, where the I_D/I_G increases from 0.50 to 0.93. This indicates a trend of an increase in the content of the sp^2−C bond with an increase in power. The higher Cr doping enhances the film's disorderliness. Moreover, increasing the power enhances the frequency of sputtering particle bombardment on the substrate surface, increasing the film surface temperature and promoting the transition of metastable sp^3−C bonds to sp^2−C bonds.

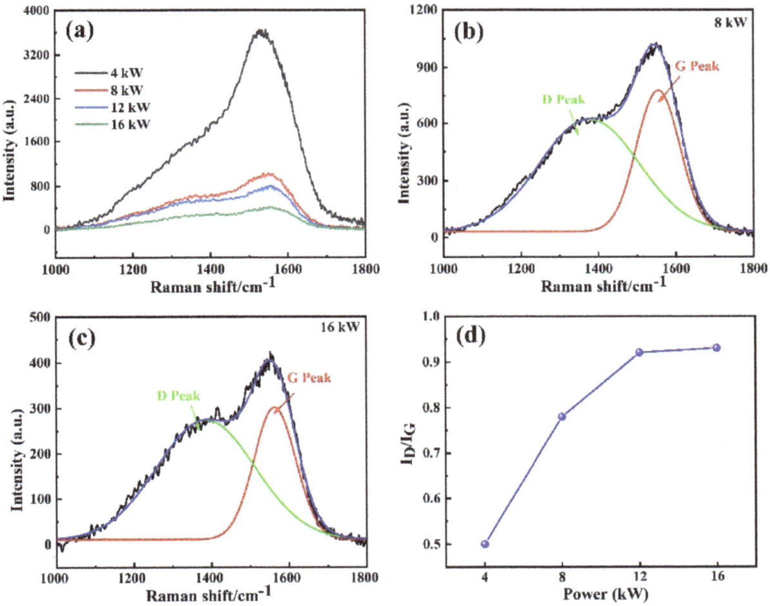

Figure 6. Representative Raman spectra of films under different target powers: (**a**) full Raman spectra, (**b**) 8 kW Raman spectrum, (**c**) 16 kW Raman spectrum, and (**d**) the corresponding I_D/I_G ratio.

To investigate the influence of target power on the chemical bonding status of carbon atoms in the films, C1s' high-resolution fitting analyses were conducted for several film groups, as shown in Figure 7. All films exhibit a main peak at around 284.6 eV and the shoulder peaks at around 283.5 eV can be observed in films deposited at 8, 12, and 16 kW. Especially, the film deposited at 16 kW shows that the intensity of the shoulder peak is comparable to that of the main peak. The C1s peak could be deconvoluted into four peaks at around 287.4 eV, 285.6 eV, 284.7 eV, and 283.1 eV, corresponding to C−O, sp^3−C, sp^2−C, and Cr−C bonds, respectively. The absence of a Cr−C peak in the film deposited at 4 kW is due to the low Cr content, as it primarily exists in a dispersed solid solution form within the amorphous carbon network. Additionally, in Figure 7d, the film displays a low binding energy peak at 282.8 eV, which was attributed to the Cr−C bond. According to previous works, the Cr−C bonds at the binding energy peaks of 282.8 eV and 283.1 eV belong to Cr_7C_3 and Cr_2C_3, respectively [21,33]. As the target power increases, the intensity of the

Cr_2C_3 peaks gradually rises, indicating that the Cr content increases in films. At the highest Cr content, as shown in Figure 7d, the Cr_7C_3 phase appears. As a carbide-forming element, the Cr doped into the film forms carbides with C. Following an increase in Cr content due to the higher Cr target power, the formation of carbides also increases. The presence of C−O bonds may be attributed to the oxygen adsorption on the sample surface in ambient air before XPS detection.

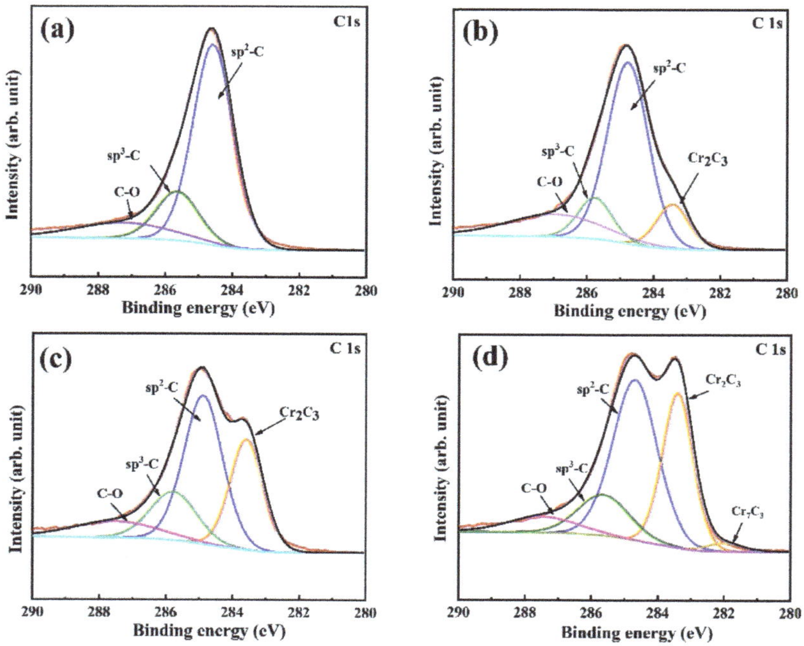

Figure 7. Typical C1s high-resolution XPS spectra of films with various target powers: (**a**) 4 kW, (**b**) 8 kW, (**c**) 12 kW, and (**d**) 16 kW.

The hardness and elastic modulus of films deposited with different powers are shown in Figure 8a. With an increase in power, the film's hardness and elastic modulus exhibit a trend of an initial increase followed by a decrease, reaching the maximum values at 12 kW, measuring at 11.44 GPa and 137.25 GPa, respectively. This is because, in cases with a small amount of doping, the metal Cr is embedded in the three-dimensional network film with different sizes of atomic clusters, which improves the hardness of the film and maintains the toughness of the film. When the Cr doping content is excessive, the strong plasticity becomes dominant, and the hardness and elastic modulus of the film decrease significantly [34]. Throughout the doping process, the elastic modulus and hardness of films demonstrate a trend of an initial increase followed by a decrease. In addition, increasing the power enhances the Cr ionization rate, elevating the Cr content in the film, and the formation of carbides reinforces the film hardness. The surface roughness (Sa) of the film exhibited a decreasing trend, corresponding to the denser structure. This also contributed to the hardness enhancement. With further increases in power, the Cr content stabilizes, but the film's surface temperature rises due to ion bombardment. This leads to graphitization, an increase in sp^2−C bonds, and a slight decrease in film hardness [35]. The characteristic coefficient H^3/E^2, representing the resistance to plastic deformation of the film, decreases with increasing power, as shown in Figure 8b. Typically, a lower H^3/E^2 value indicates a poorer wear resistance of the film [36]. At 4 kw, the plastic factor value is the highest, indicating better resistance to plastic deformation, while at a 16 kW deposition power, the plastic factor value is lowest, indicating poorer resistance to plastic deformation.

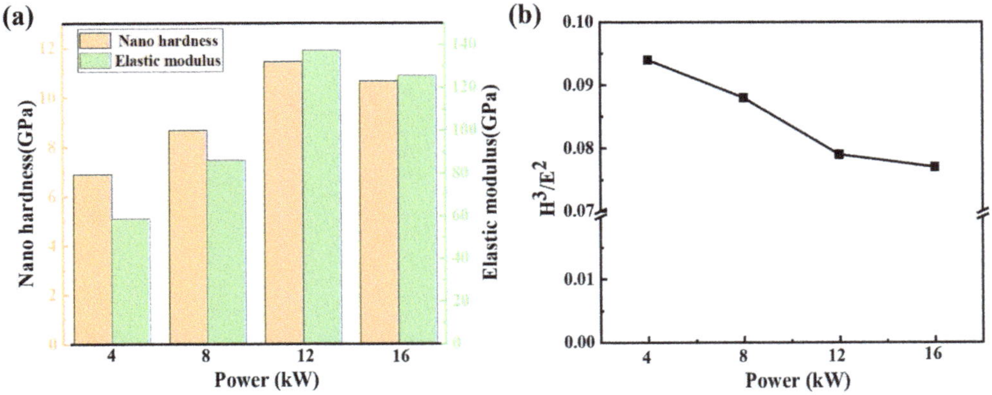

Figure 8. Mechanical properties of films at different target powers: (**a**) hardness and elastic modulus; (**b**) the ratio of H^3/E^2.

The variation in the friction coefficient of films under different target powers is depicted in Figure 9a. The films deposited at relatively low target powers (4 kW and 8 kW) enter a stable friction coefficient (COF) stage after a brief running-in period. At the stable stage, the film deposited at 4 kW showed a COF at around 0.17 and that of the film deposited at 8 kW remained around 0.14. The COFs of the films deposited at high target powers exhibit significant fluctuations, with higher values at the onset of friction. The COF of the film deposited at 16 kW reached 0.5. This is because when the ball and film first come into contact, the ball surface does not form a lubricating transfer film, resulting in higher friction coefficient values within the initial 10,000 cycles. As the number of friction cycles increases, the films gradually enter a stable running-in period, and after 15,000 cycles, the friction coefficients stabilize at around 0.25 and 0.31, respectively. The corresponding wear rates of films are shown in Figure 9c. As the power increases from 4 kW to 16 kW, the wear rate of the film rises from 1.28×10^{-7} mm^3/N·m to 2.48×10^{-6} mm^3/N·m. This is attributed to the presence of hard carbide phases in the film at high doping levels, which cause abrasive wear and result in greater wear depths, leading to an increase in the wear rate [37]. The changing trend of wear rate described above is also consistent with the result of H^3/E^2, shown in Figure 8b.

Figure 10 displays the 2D profile curves of films after the wearing test. At 4 kW power, the film exhibits the shallowest wear scar, less than 0.4 μm deep, and the wear width is around 220 μm. The corresponding wear rate is the lowest, as shown in Figure 9c, indicating superior wear resistance. As the target power increases from 8 kW to 16 kW, the width of the wear scar increases from 188 μm to 430 μm, and the depth increases from 0.5 μm to 2 μm. This is attributed to the higher Cr content resulting in the formation of a large quantity of hard carbides, which act as wear debris, leading to an increase in the friction coefficient. To investigate the reasons for the different performance of films prepared at different powers under the same friction experiment conditions, the wear scars were analyzed using Raman spectroscopy after the friction experiment. The results, as shown in Figure 11, indicate that the I_D/I_G values of the wear scars for all films are higher than the I_D/I_G values of the films themselves, suggesting the graphitization of the transfer film during friction. This indicates that during high-speed friction between the ball and the film, the generated heat and the interaction forces cause the sp^3 bonds within the film to transition to sp^2 bonds. Previous work also proved that the sp^3−C atoms could only be short-lived and gradually transform into sp^2−C atoms with the graphitization generated on the wear surface of the films [38]. A graphite transfer film forms on the surface of the ball, providing lubrication during the friction process. Therefore, with an increase in friction time, the friction coefficients of the films decrease.

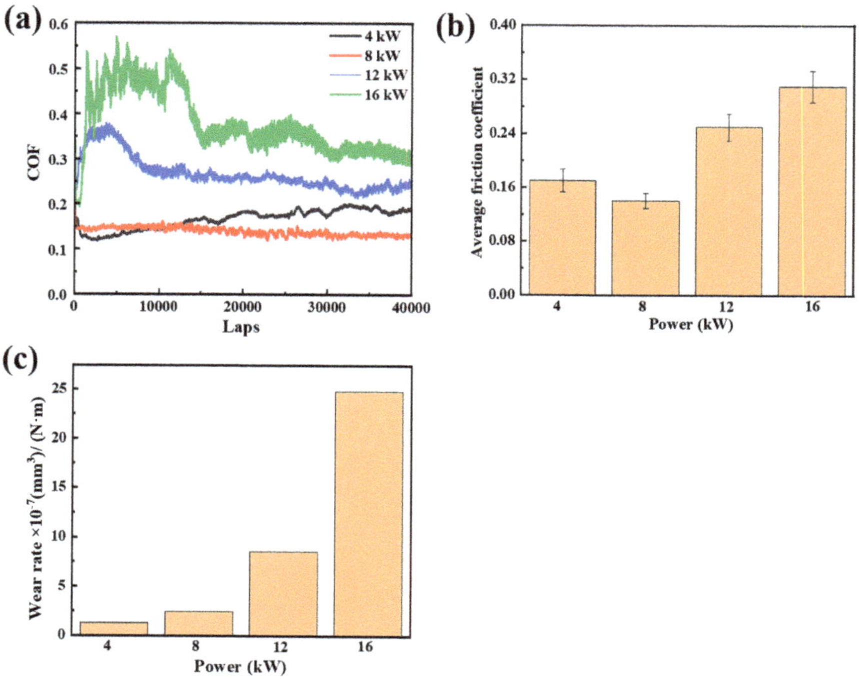

Figure 9. (**a**) Friction coefficient curves, (**b**) average friction coefficient, and (**c**) wear rates of films with various target powers.

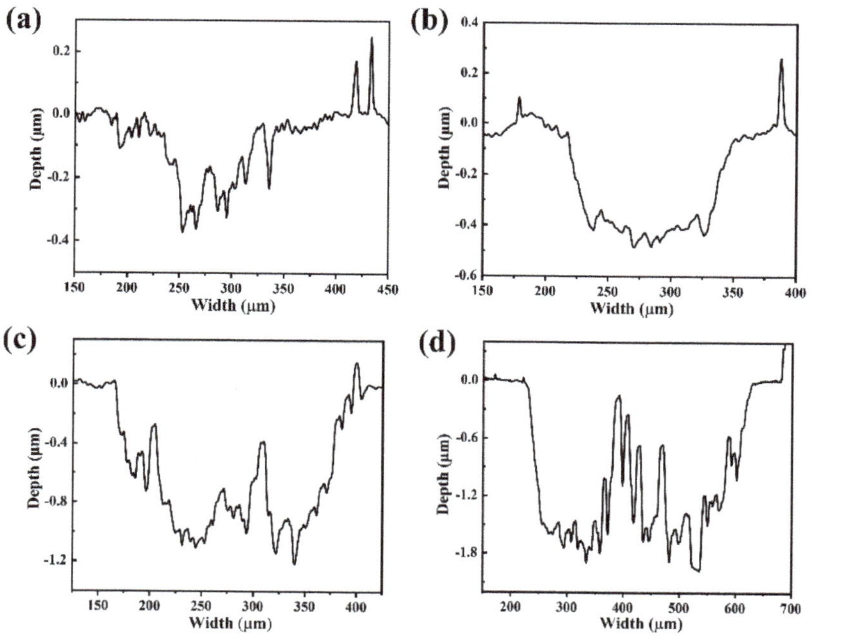

Figure 10. Wear profile curves of films with various target powers: (**a**) 4 kW, (**b**) 8 kW, (**c**) 12 kW, and (**d**) 16 kW.

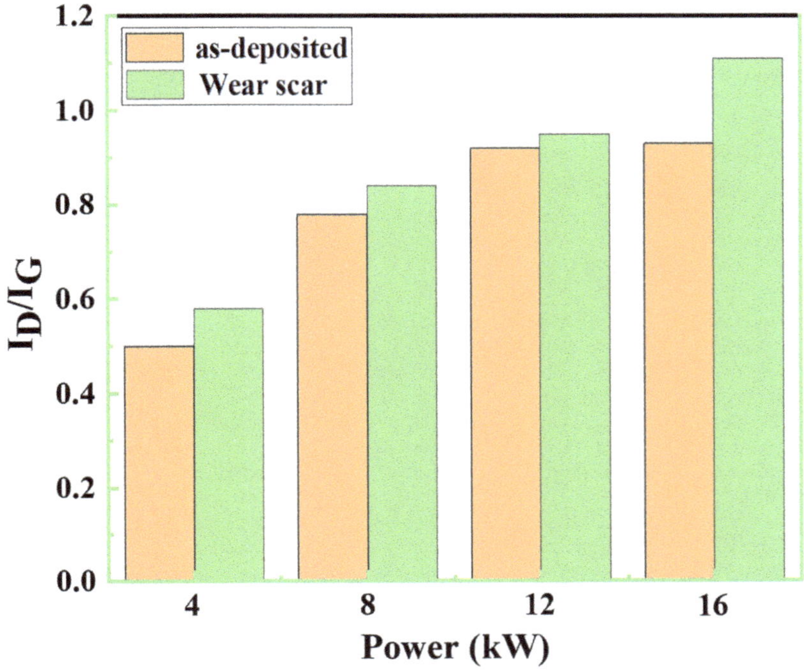

Figure 11. I_D/I_G values of films and wear scars.

4. Conclusions

In this work, to improve the comprehensive properties of titanium alloys, diamond-like carbon films with Cr doping on the surfaces of titanium alloys were deposited via high-power pulsed magnetron sputtering technology. The considerable properties of titanium alloy substrates, including mechanical and tribological properties, were achieved. The effect of target power on the microstructure and the mechanical and tribological properties of the films was studied in detail. It was found that the Cr content in films could be adjusted by varying the Cr target power, increasing from 3.73 at. % to 22.65 at. % with an increase in target power. The microstructure of the film evolved from an amorphous structure to the nanocomposite structure consisting of carbides embedded in an amorphous carbon matrix. The I_D/I_G ratios linearly increased, implying that the sp^2-C content in the film could be governed by changing the target power. The hardness of the films increased with an increase in power from 4 to 12 kW, and decreased slightly with a power of 16 kW. The formed hard carbides were responsible for the hardness enhancement. Serious abrasive wear occurred in the films deposited at a higher target power due to the excessive hard particles. The wear rates showed the same changing trend as H^3/E^2 as the target power rose from 4 kW to 16 kW. The lowest wear rate of 1.28×10^{-7} $mm^3/N{\cdot}m$ was obtained for the film deposited at 4 kW.

Author Contributions: Writing—original draft, S.L.; investigation, W.Z.; writing—review and editing and supervision, J.D.; methodology, Y.L.; formal analysis, W.Y.; conceptualization and data curation, Y.Y.; software, X.L.; formal analysis and resources, J.Y.; project administration, J.Z. All authors have read and agreed to the published version of the manuscript.

Funding: This work was financially supported by the National Natural Science Foundation of China (No. 52101063), the Natural Science Foundation of the Anhui Higher Education Institutions of China (No. 2022AH050312; KJ2021A0392), and the Open Project of China International Science and Technology Cooperation Base on Intelligent Equipment Manufacturing in Special Service Environment (No. ISTC2023KF02).

Institutional Review Board Statement: Not applicable.

Informed Consent Statement: Not applicable.

Data Availability Statement: Data are contained within the article.

Conflicts of Interest: Author Weibo Yu was employed by the company Changzhou Huachuang Aviation Technology Co., Ltd. Author Jing Yuan was employed by the company Anhui HERO Electronic Sci & Tec Co., Ltd.

References

1. Sadeghpour, S.; Javaheri, V.; Bruschi, S.; Kömi, J.; Karjalainen, P. Strain rate and mechanical stability in determining deformation behavior of beta Ti alloys. *Mater. Sci. Eng. A* **2020**, *798*, 140274. [CrossRef]
2. Ren, L.; Xiao, W.; Kent, D.; Wan, M.; Ma, C.; Zhou, L. Simultaneously enhanced strength and ductility in a metastable β-Ti alloy by stress-induced hierarchical twin structure. *Scr. Mater.* **2020**, *184*, 6–11. [CrossRef]
3. Shao, L.; Li, W.; Li, D.; Xie, G.; Zhang, C.; Zhang, C.; Huang, J. A review on combustion behavior and mechanism of Ti alloys for advanced aero-engine. *J. Alloys Compd.* **2023**, *960*, 170584. [CrossRef]
4. Jain, S.; Parashar, V. Comparison of priori and posteriori approach of multi-objective optimization for wedm on Ti6Al4V alloy. *Mater. Res. Express* **2022**, *9*, 076504. [CrossRef]
5. Zhang, B.; Cai, Z.; Gan, X.; Zhu, M.; Yu, H. Dual motion fretting wear behaviors of titanium and its alloy in artificial saliva. *Trans. Nonferr. Met. Soc. China* **2014**, *24*, 100–107. [CrossRef]
6. Kashyap, V.; Ramkumar, P. DLC coating over pre-oxidized and textured Ti6Al4V for superior adhesion and tribo-performance of hip implant. *Surf. Coat. Technol.* **2022**, *440*, 128492. [CrossRef]
7. Ghahramanzadeh Asl, H. Investigation of friction and wear performance on oxidized Ti6Al4V alloy at different temperatures by plasma oxidation method under ambient air and vacuum conditions. *Vacuum* **2020**, *180*, 109578. [CrossRef]
8. Zhou, Z.-Y.; Liu, X.-B.; Zhuang, S.-G.; Yang, X.-H.; Wang, M.; Sun, C.-F. Preparation and high temperature tribological properties of laser in-situ synthesized self-lubricating composite coatings containing metal sulfides on Ti6Al4V alloy. *Appl. Surf. Sci.* **2019**, *481*, 209–218. [CrossRef]
9. Kümmel, D.; Schneider, J.; Gumbsch, P. Influence of interstitial oxygen on the tribology of ti6al4v. *Tribol. Lett.* **2020**, *68*, 96. [CrossRef]
10. Wang, J.; Ma, J.; Huang, W.; Wang, L.; He, H.; Liu, C. The investigation of the structures and tribological properties of f-dlc coatings deposited on Ti-6Al-4V alloys. *Surf. Coat. Technol.* **2017**, *316*, 22–29. [CrossRef]
11. Zhang, S.; Huang, T.; Sun, S.; Wu, S.; Yang, X.; Guo, F.; Zhang, B.; Dai, L. Effects of bias voltages on the tribological behaviors of DLC Coatings. *Coatings* **2024**, *14*, 176. [CrossRef]
12. Wang, J.; Wang, L.; Chen, H.; Wang, H. Molecular dynamics simulation of friction in dlc films with different sp3 contents. *Tribol. Int.* **2023**, *190*, 109050. [CrossRef]
13. Tillmann, W.; Wittig, A.; Dias, N.F.L.; Stangier, D.; Thomann, C.A.; Moldenhauer, H.; Debus, J. Silicon- and tungsten-containing hydrogen-free and hydrogenated amorphous carbon films for friction-reducing applications. *Diam. Relat. Mater.* **2022**, *123*, 108866. [CrossRef]
14. Sha, B.; Lukianov, A.N.; Dusheiko, M.G.; Lozinskii, V.B.; Klyui, A.N.; Korbutyak, D.V.; Pritchin, S.E.; Klyui Nickolai, I. Carbon-rich amorphous silicon carbide and silicon carbonitride films for silicon-based photoelectric devices and optical elements: Application from uv to mid-ir spectral range. *Opt. Mater.* **2020**, *106*, 109959. [CrossRef]
15. Li, X.; Hou, K.; Qiu, D.; Yi, P.; Lai, X. A first principles and experimental study on the influence of nitrogen doping on the performance of amorphous carbon films for proton exchange membrane fuel cells. *Carbon* **2020**, *167*, 219–229. [CrossRef]
16. Alotaibi, S.; Nama Manjunatha, K.; Paul, S. Stability of hydrogenated amorphous carbon thin films for application in electronic devices. *Diam. Relat. Mater.* **2018**, *90*, 172–180. [CrossRef]
17. Wu, J.; Wei, P.; Liu, G.; Chen, D.; Zhang, X.; Chen, T.; Liu, H. A comprehensive evaluation of DLC coating on gear bending fatigue, contact fatigue, and scuffing performance. *Wear* **2024**, *536–537*, 205177. [CrossRef]
18. Ding, J.C.; Cheng, Y.; Zhang, S.; Wang, Q.; Zhang, T.F. Microstructure, mechanical and tribological properties of ti doped ta-c films deposited by a hybrid coating system. *Diam. Relat. Mater.* **2023**, *131*, 109565. [CrossRef]
19. Ding, J.C.; Dai, W.; Zhang, T.F.; Zhao, P.; Yun, J.M.; Kim, K.H.; Wang, Q.M. Microstructure and properties of nb-doped diamond-like carbon films deposited by high power impulse magnetron sputtering. *Thin Solid Films* **2018**, *663*, 159–167. [CrossRef]
20. Guo, C.-Q.; Li, H.-Q.; Peng, Y.-L.; Dai, M.-J.; Lin, S.-S.; Shi, Q.; Wei, C.-B. Residual stress and tribological behavior of hydrogen-free al-dlc films prepared by hipims under different bias voltages. *Surf. Coat. Technol.* **2022**, *445*, 128713. [CrossRef]
21. Santiago, J.A.; Fernández-Martínez, I.; Sánchez-López, J.C.; Rojas, T.C.; Wennberg, A.; Bellido-González, V.; Molina-Aldareguia, J.M.; Monclús, M.A.; González-Arrabal, R. Tribomechanical properties of hard cr-doped dlc coatings deposited by low-frequency hipims. *Surf. Coat. Technol.* **2020**, *382*, 124899. [CrossRef]

22. Jain, V.; Roychowdhury, T.; Kuimelis, R.G.; Linford, M.R. Differences in surface reactivity in two synthetic routes between hipims and dc magnetron sputtered carbon. *Surf. Coat. Technol.* **2019**, *378*, 125003. [CrossRef]
23. Ma, D.; Harvey, T.J.; Wellman, R.G.; Ehiasarian, A.P.; Hovsepian, P.E.; Sugumaran, A.A.; Purandare, Y.P.; Wood, R.J.K. Cavitation erosion performance of cralyn/crn nanoscale multilayer coatings deposited on Ti6Al4V by hipims. *J. Alloys Compd.* **2019**, *788*, 719–728. [CrossRef]
24. Costa, A.; Ferreira, F.; Colaux, J.L.; Vahidi, A.; Serra, R.; Oliveira, J. Effect of hydrogen incorporation on the mechanical properties of dlc films deposited by hipims in doms mode. *Surf. Coat. Technol.* **2023**, *473*, 129980. [CrossRef]
25. Solomon, I.; Bhatnagar, M.; Shukla, K.; Sarma, B.; Ranjan, M.; Sarma, A. Correlation of structural and optical properties of pvd grown amorphous carbon thin films. *Diam. Relat. Mater.* **2017**, *75*, 69–77. [CrossRef]
26. Kouznetsov, V.; Macák, K.; Schneider, J.M.; Helmersson, U.; Petrov, I. A novel pulsed magnetron sputter technique utilizing very high target power densities. *Surf. Coat. Technol.* **1999**, *122*, 290–293. [CrossRef]
27. Ganesan, R.; Fernandez-Martinez, I.; Akhavan, B.; Matthews, D.T.A.; Sergachev, D.; Stueber, M.; McKenzie, D.R.; Bilek, M.M.M. Pulse length selection in bipolar hipims for high deposition rate of smooth, hard amorphous carbon films. *Surf. Coat. Technol.* **2023**, *454*, 129199. [CrossRef]
28. Brenning, N.; Hajihoseini, H.; Rudolph, M.; Raadu, M.A.; Gudmundsson, J.T.; Minea, T.M.; Lundin, D. HiPIMS optimization by using mixed high-power and low-power pulsing. *Plasma Sources Sci. Technol.* **2021**, *30*, 015015. [CrossRef]
29. Huo, C.; Lundin, D.; Raadu, M.A.; Anders, A.; Gudmundsson, J.T.; Brenning, N. On the road to self-sputtering in high power impulse magnetron sputtering: Particle balance and discharge characteristics. *Plasma Sources Sci. Technol.* **2014**, *23*, 025017. [CrossRef]
30. Fernandes, F.; Calderon, V.S.; Ferreira, P.J.; Cavaleiro, A.; Oliveira, J.C. Low peak power deposition regime in hipims: Deposition of hard and dense nanocomposite Ti-Si-n films by doms without the need of energetic bombardment. *Surf. Coat. Technol.* **2020**, *397*, 125996. [CrossRef]
31. Adelhelm, C.; Balden, M.; Rinke, M.; Stueber, M. Influence of doping (Ti, V, Zr, W) and annealing on the sp^2 carbon structure of amorphous carbon films. *J. Appl. Phys.* **2009**, *105*, 033522. [CrossRef]
32. Peng, Y.-L.; Guo, C.-Q.; Lin, S.-S.; Shi, Q.; Wei, C.-B.; Su, Y.-F.; Wu, Y.-Q.; Tang, P.; Zhu, X.-G.; Dai, M.-J. Effects of working pressure on structure and properties of al-containing amorphous carbon films prepared by high-power impulse magnetron sputtering. *J. Alloys Compd.* **2020**, *816*, 152587. [CrossRef]
33. NIST X-ray Photoelectron Spectroscopy Database (SRD 20), Version 5.0. Available online: https://srdata.nist.gov/xps/EnergyTypeElement (accessed on 25 June 2024).
34. Zhu, L.; Li, J.; Kang, J.; Tang, L.; Ma, G.; Han, C.; Shi, J.; Wang, H. Different cr contents on the microstructure and tribomechanical properties of multi-layered diamond-like carbon films prepared by unbalanced magnetron sputtering. *J. Mater. Eng. Perform.* **2020**, *29*, 7131–7140. [CrossRef]
35. Yang, L.; Chen, Y.; Xu, Z.; Toshiaki, N.; Xi, Y.; Ni, Q. Effect of heat treatment on mechanical property of amorphous carbon films by magnetron sputtering. *Diam. Relat. Mater.* **2022**, *129*, 109328. [CrossRef]
36. Tao, H.; Tsai, M.T.; Chen, H.-W.; Huang, J.C.; Duh, J.-G. Improving high-temperature tribological characteristics on nanocomposite cralsin coating by mo doping. *Surf. Coat. Technol.* **2018**, *349*, 752–756. [CrossRef]
37. Dai, W.; Ke, P.; Wang, A. Microstructure and property evolution of cr-dlc films with different Cr content deposited by a hybrid beam technique. *Vacuum* **2011**, *85*, 792–797. [CrossRef]
38. Zhou, Y.; Chen, Z.; Zhang, T.; Zhang, S.; Xing, X.; Yang, Q.; Li, D. Metastable hybridized structure transformation in amorphous carbon films during friction—A study combining experiments and MD simulation. *Friction* **2023**, *11*, 1708–1723. [CrossRef]

Disclaimer/Publisher's Note: The statements, opinions and data contained in all publications are solely those of the individual author(s) and contributor(s) and not of MDPI and/or the editor(s). MDPI and/or the editor(s) disclaim responsibility for any injury to people or property resulting from any ideas, methods, instructions or products referred to in the content.

coatings

MDPI

Article

Development and Mechanical Characterization of Ni-Cr Alloy Foam Using Ultrasonic-Assisted Electroplating Coating Technique

Raj Kumar Pittala [1], Priyaranjan Sharma [1,*], Gajanan Anne [2], Sachinkumar Patil [3], Vinay Varghese [4], Sudhansu Ranjan Das [5,*], Ch Sateesh Kumar [6,7,*] and Filipe Fernandes [8,9]

1 Department of Mechanical Engineering, Koneru Lakshmaih Education Foundation, Vaddeswaram 522502, Andhra Pradesh, India
2 Department of Mechanical and Industrial Engineering, Manipal Institute of Technology, Manipal Academy of Higher Education, Manipal 576104, Karnataka, India; gajanan.anne25@gmail.com
3 Department of Mechanical Engineering, REVA University, Bangalore 560064, Karnataka, India; sachinkumar.patil@reva.edu.in
4 Department of Mechanical Engineering, Adishankara Institute of Engineering & Technology, Cochin 683574, Kalady, India; vinujavs@gmail.com
5 Department of Production Engineering, Veer Surendra Sai University of Technology, Burla 768018, Odisha, India
6 CFAA-Aeronautics Advanced Manufacturing Center, University of the Basque Country (UPV/EHU), Biscay Science and Technology Park, Ed. 202, 48940 Zamudio, Spain
7 Department of Mechanical Engineering, University of the Basque Country, Escuela Superior de Ingenieros Alameda de Urquijo S/N, 48013 Bilbao, Spain
8 Department of Mechanical Engineering, ARISE, CEMMPRE, University of Coimbra, Rua Luís Reis Santos, 3030-788 Coimbra, Portugal; filipe.fernandes@dem.uc.pt
9 ISEP, Polytechnic of Porto, Rua Dr. António Bernardino de Almeida, 4249-015 Porto, Portugal
* Correspondence: priya333ranjan@gmail.com (P.S.); das.sudhansu83@gmail.com (S.R.D.); chigullasateesh.kumar@ehu.eus (C.S.K.)

Abstract: Metal foams and alloy foams are a novel class of engineering materials and have numerous applications because of their properties such as high energy absorption, light weight and high compressive strength. In the present study, the methodology adopted to develop a Ni-Cr alloy foam is discussed. Polyurethane (PU) foam of 40PPI (parts per inch) pore density was used as the precursor and coating techniques such as electroless nickel plating (ELN), ultrasonic-assisted electroplating of nickel (UAEPN), and pack cementation or chromizing were used to develop the Ni-Cr alloy foam. The surface morphology, strut thickness and minimum weight gain after each coating stage were evaluated. It was observed from the results that the adopted coating techniques did not damage the original ligament cross-section of the PU precursor. The minimum weight gain and the coating thickness after the UAEPN process were observed to be 42 g and 40–60 μm, respectively. The properties such as porosity percentage, permeability and compressive strength were evaluated. Finally, the pressure drop through the developed foam was estimated and verified to determine whether the developed foam can be used for filtering applications.

Keywords: Ni-Cr alloy foam; electroless nickel plating; ultrasonic-assisted; chromizing; porosity

1. Introduction

Porous materials such as alloy foams and metal foams have abundant engineering applications due to their exceptional physical and mechanical properties. Basically, an alloy or metal foam is a cellular structure that has a solid matrix made of an alloy or metal with empty voids. If the voids are connected by means of open pores, then the foam is described as an open-cell foam, and if the voids are not connected by open channels and are disconnected by solid walls then the foam is said to be a closed-cell foam. Alloy foams offer significant performance gain in light and stiff structures for high energy absorption, better thermal management and high compressive strength [1]. These foams are reproducible and non-toxic. Thus, foams are one of the best lightweight materials that can be employed in many functional and structural applications.

Most commercially existing metal foams are based on copper, nickel, zinc, aluminum, titanium and their alloys. Aluminum alloy foams are mainly used in sound absorption, thermal insulation and vibration control applications [2]. Nickel-based alloy foams such as Ni-Cr foams combine good mechanical properties and better corrosion resistance at room temperature as well as for high-temperature applications. Because of their low density, high specific area and good ductility, Ni-Cr alloy foams can be used as sandwich-structure core material and for mechanical damping applications [3]. These foams are also used to produce near-net-shape complex components, and for filtering and separation applications, due to their unique microstructure. Chao-Heng et al. [4] investigated a nickel foam for photocatalytic filtering applications and concluded that the removal of formaldehyde from indoor air is made better by incorporating the nickel foam in a standard air filter.

Several techniques such as slurry foaming [5,6], direct foaming with a blowing agent [7,8], investment casting [9,10] and electrochemical deposition [11–13] are available to fabricate alloy foams. The microstructure and mechanical properties of the foam greatly depends on the fabrication technique. To use the developed foam for filtering and separation applications, maintaining the original porosity, ligament or strut structure is necessary for better functional performance. Hence, in the present study, electrochemical deposition techniques were used to develop Ni-Cr alloy foams. Duan et al. [14] developed a Ni-Cr alloy foam using electrodeposition techniques and investigated the various process parameters for optimization of the microstructure and higher electrical resistivity. Pang Qiu et al. [15] synthesized Ni-Fe-Cr alloy foams employing gas-phase co-deposition techniques and studied their compressive and energy absorption properties. They concluded that both properties were increased with an increase in Cr content in the developed foam.

The mechanical properties of the cellular structures depend on the pore size, relative density, strut material and porosity percentage. Michailidis et al. [16] fabricated an open-cell aluminum foam using a space holder technique and evaluated its compressive strength. It was understood from their study that the stress–strain curve of the foam depends on the pore size and that the compressive strength reduces with enhancement of the porosity percentage. Kaplon et al. [17] used two different polyurethane (PU) foam materials as precursor materials and developed open-porosity magnesium foam using an investment-casting technique, and they examined its microstructure, corrosion resistance and mechanical compressive strength. Zhang et al. [18] developed geopolymer foams by employing aluminum and H_2O_2-sodium oleate as foaming agent and optimized the pore structure. Chemical compositions and microstructure were analyzed using Fourier transform infrared spectroscopy and X-ray diffraction (XRD) techniques and they measured the permeability and flow properties using a permeameter.

From the literature survey, it can be summarized that a few studies [17,19,20] have used aluminum, zinc, magnesium and copper metals as foams, while some other studies [14,15,21] have used nickel alloys, aluminum alloys and titanium-based copper alloys as foams. In a few works [7,8,10], melt foaming, direct foaming with a blowing agent, and investment casting were used to fabricate metal or alloy foams. In a few other works [19,22–24], electrodeposition techniques and powder metallurgy methods were used to fabricate alloy foams. Some studies [25–27] investigated the microstructure, mechanical properties and process parameters of the developed foams, and some other studies [28,29] concentrated on the functional performance and optimum pore-size structure. Overall, nickel-based and aluminum-based open-cell foams have been discussed without much focus on the fluid flow characteristics and permeability calculations.

Because of new advancements in the aerospace, defense and automotive industries, there is a large demand for lightweight structures which can be used as filtering media, as well as structural components to increase fuel efficiency and functional performance. Thus, to satisfy these requirements, the development methodology for such a novel structural material, a Ni-Cr alloy foam, is discussed in the present study. As per the knowledge of the authors, there is no study available on Ni-Cr foams which addresses their process capability, pressure drop and permeability calculations. Very limited studies are available on the synthesis of Ni-Cr alloy foams using electrodeposition techniques. Thus, the objectives of the present study are (i) to discuss the methodology adopted to develop a Ni-Cr alloy foam and its process parameters' optimization; (ii) to evaluate the process capability of the various methods used to develop such a foam; and (iii) to determine its porosity percentage, permeability, pressure drop and mechanical compressive strength.

2. Materials and Methods

2.1. Materials

Polyurethane foam having 40 PPI (parts per inch) pore density and a strut thickness of 100–110 µm was taken as precursor and purchased from Sheela foam Ltd., Noida, India. All the reagents used in the electroless nickel plating (ELN), ultrasonic-assisted electroplating of nickel (UAEPN) and pack chromizing were of analytical grade and purchased from Alfa Aesar, Mumbai, India. Demineralized (DM) water was used to prepare different bath compositions.

2.2. Synthesis of Ni-Cr Alloy Foam

Figure 1 shows a schematic representation of the Ni-Cr alloy foam development methodology. The methodology adopted to develop the Ni-Cr foam was mainly electrodeposition techniques and solid-state diffusion techniques. Initially, PU foam having 60 × 60 × 15 mm^3 dimensions and 40 PPI was taken as a base material and metallized using an electroless nickel-plating technique. The detailed bath composition and the ELN procedure are described in Figure 2. As an output of the ELN process, a thin layer of Ni atoms was deposited on the precursor foam. The optimized process parameters were temperature: 90–95 °C, time: 50–60 min and pH: 4.5–5.5. In order to achieve uniform deposition and the required thickness of Ni coating on the metallized foam, an ultrasonic-assisted electroplating of nickel (UAEPN) technique was used. Table 1 represents the bath composition and operating parameters used in the study. The process scheme of the UAEPN technique is shown in Figure 3. The ELN-coated Ni foam was used as cathode, while two nickel plates of size 80 × 80 × 15 mm^3 were used as anodes during the UAEPN process. Later, in order to remove the PU precursor from the developed Ni foam, sintering was carried out at 1100 °C for 1 h. A resistance-heat furnace of 9 kW capacity and reducing atmosphere (H_2) was used during sintering. In the sintering process, the PU precursor was burnt out at 450–500 °C and strengthening of the EPN foam was carried out by creating bonding between the plated Ni particles.

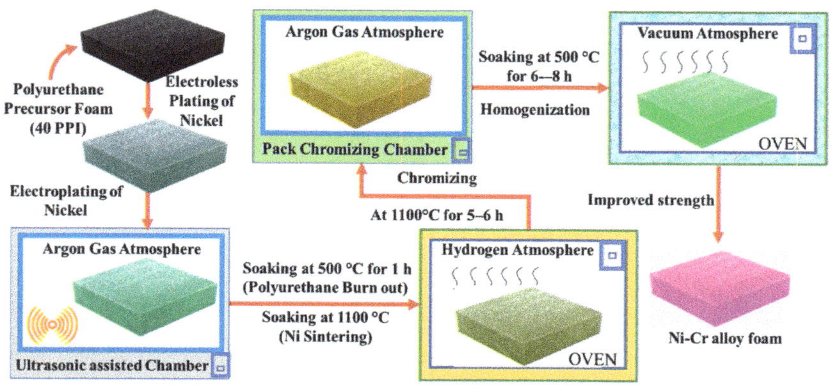

Figure 1. Schematic illustration of Ni-Cr alloy foam development methodology.

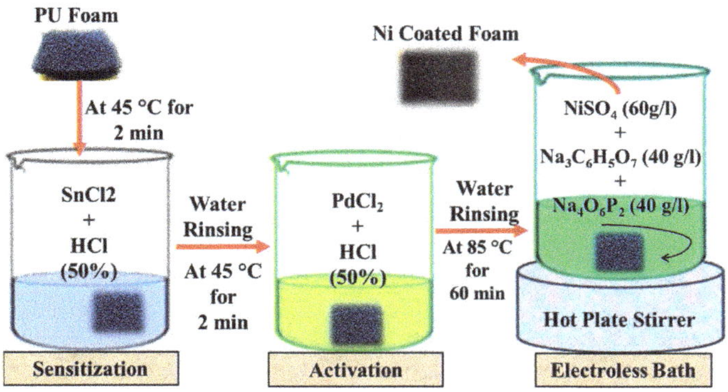

Figure 2. Process scheme of ELN technique.

Table 1. Bath composition and process parameters of UAEPN technique.

Process	Bath Composition	Operating Parameters
UAEPN	Ni salt ($NiSO_4 \cdot 6H_2O$ + $NiCl_2 \cdot 6H_2O$): 350 g/L Boric acid: 20 g/L Sodium sulphate: 15 g/L Omni additive 992: 8 mL/L Magnum brightener 437: 10 mL/L	Current: 7–8.5 A Voltage: 10–12 V pH: 4–5 Duration: 300–360 min

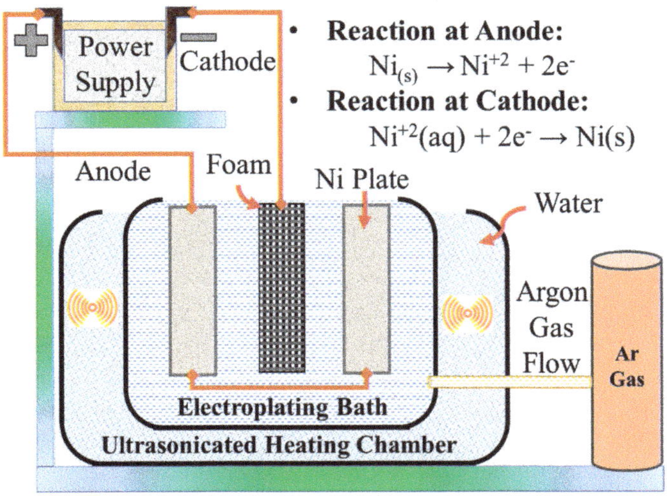

Figure 3. Process scheme of UAEPN technique.

To deposit the chromium on the sintered Ni foam, a chromizing or pack-cementation process was used. Chromizing is a solid-state diffusion process in which diffusion happens through the difference in concentrations of the depositing material. To start with the chromizing process, first, ball milling of the chrome pack was carried out. The chrome pack consisted of 55 wt.% Cr powder, 15 wt.% NH_4Cl (activator) and 30 wt.% Al_2O_3 (filler material). The powders were carefully mixed using a mortar and pestle and underwent high-energy ball milling for 5 h. Ball milling of the pack ensured uniform mixing (blending) of powders and reduction in the powder particle size. Then, the entire pack composition was collected and kept in an SS tray of size 65 × 70 × 15 mm^3. Then, the tray was loaded with sintered Ni foam and the pack composition and loaded in the resistance furnace and allowed to heat up to 1100 °C at a heating rate of 8 °C/min. Argon was used as a shielding gas to avoid oxidation of the foam during chromizing. The sample was soaked for 6 h at 1100 °C and then allowed for furnace cooling.

2.3. Characterization Techniques

Different characterization techniques such as scanning electron microscopy (SEM), X-ray diffraction (XRD), electron dispersive spectroscopy (EDS) (TESCAN ORSAY HOLDING, Brno, Czech Republic) and compression tests were performed to evaluate the microstructure and mechanical behavior of the developed alloy foam.

Strut thickness, morphology and elemental mapping of the developed foam were examined using Vega 3 TESCAN SEM (TESCAN ORSAY HOLDING, Brno, Czech Republic) at a working depth of 10–11 mm and 10 kV voltage.

In order to know the phase formations and alloys formed between the Ni and Cr, XRD (Philips, Caerphilly, UK) analysis was used. The developed foam and a Ni 60 Cr 40 solid sample were compared using XRD to confirm the bonding between the Ni and Cr elements.

The quasi-static compression behavior and the maximum compressive strength of the developed foam were examined using a 50 kN load cell Universal Testing Machine (UTM—Jin Ahn Testing, Suzhou, China). A cylindrical specimen of 12 mm diameter and 15 mm height was considered and tested according to the ASTM C365-05 standard. The lowest possible dimension of the sample was selected, such that it should be at least seven times the cell size to avoid size effects. Specimens were kept on the bottom ram of the UTM machine and load was applied to the specimens when the top ram moved downwards at a constant cross-head speed of 0.5 mm/min.

2.3.1. Measurement of Theoretical Porosity

The theoretical porosity of the developed foam was measured by using the water displacement method and Equation (1).

$$\text{Porosity}(\%) = \frac{\text{Pores volume } (S - D)}{\text{Bulk volume } (S - I)} \tag{1}$$

where S, D and I indicate the weight of a saturated sample, a dry sample and a sample immersed in water, respectively.

The weight of a saturated sample indicates the weight of the foam when its pores are sealed with water. Thus, the weight of water in pore space = S − D and the weight of water displaced = S − I.

$$\text{Pores Volume } (V_p) = \frac{w_{\text{Water}}(S - D)}{\text{density}_{\text{water}}} \tag{2}$$

If we consider the density of water to be 1 g/cc, then the pores volume will be equal to the weight of water in pore space. Similarly, the bulk volume can be expressed in terms of the weight of water displaced.

2.3.2. Measurement of Permeability and Pressure Drop through Foam

The developed foam can have application as an oil–air separator; hence, to support its functionality, the measurement of its permeability and pressure drop through it is necessary. Permeability is a property of porous materials which indicates the ease of flow of fluids through them. Since the porosity of foam varies at the different processing stages of the foam, it is also necessary to evaluate permeability at those stages.

Tadrist et al. [30] derived the relationship between the pressure drop (dp), the porosity (ε) of the medium and an average particle diameter (d_P) using Equation (3).

$$\frac{dP}{dX} = A\frac{(1-\varepsilon)^2}{\varepsilon^3 {d_P}^2}\mu V + B\frac{(1-\varepsilon)}{\varepsilon^3 d_P}\rho V^2 \tag{3}$$

A and B are the constants which differ according to the porous medium. For the Ni-Cr foam porous medium, the values of A and B can be taken as 100 and 1.0, respectively. μ, V indicates dynamic viscosity and velocity of the fluid, dX represents thickness of the foam and ρ-density of the fluid.

The relationship between particle diameter (d_p) and pore size (d) of the foam is given by Equation (4).

$$d_p = 1.5\frac{1-\varepsilon}{\varepsilon}d \quad (4)$$

The relationship between permeability (K), theoretical porosity (ε) and particle diameter (d_p) is given by Equation (5).

$$K = \frac{{d_p}^2\varepsilon^3}{A(1-\varepsilon)^2} \quad (5)$$

3. Results and Discussion

3.1. Surface Morphology and Process Capability of ELN- and UAEPN-Coated PU Foams

In order to examine the ligament cross-section of the foam and to evaluate the coating thickness, SEM analysis was used. Figure 4 shows the surface morphology and the strut thickness of the ELN- and UAEPN-coated foams, and the process capability of the ELN and UAEPN processes. It can be noted from the results that ELN-coated foam has strut thickness in the range of 115–125 μm and that of the UAEPN-coated foam is in the range of 160–180 μm. As the strut thickness of the PU precursor was 100–110 μm, these morphology results indicate that the thickness of the coating after the ELN process was around 15–25 μm and after the UAEPN process it was around 40–60 μm. To evaluate the process capability of the ELN and UAEPN processes, 10 different foam samples of the same size were considered and coated using the same process parameters. It was observed from the results that the minimum weight gain after the ELN process was around 2 g and after the UAEPN process it was around 42 g. Thus, from this study, it can be concluded that the selected process parameters can be treated as the ideal parameters.

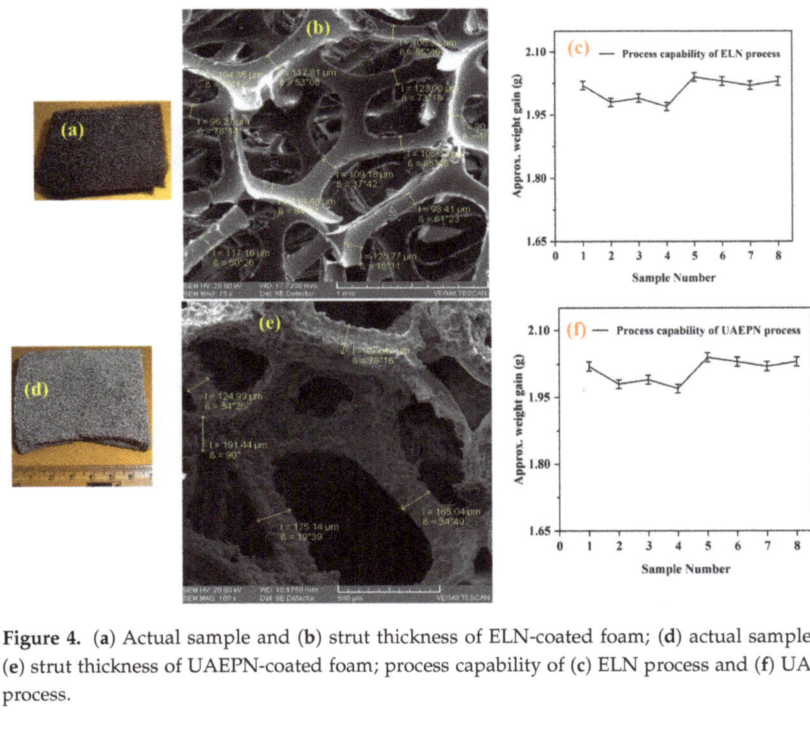

Figure 4. (**a**) Actual sample and (**b**) strut thickness of ELN-coated foam; (**d**) actual sample and (**e**) strut thickness of UAEPN-coated foam; process capability of (**c**) ELN process and (**f**) UAEPN process.

3.2. Surface Morphology and Process Capability of Chromized Foam

In order to confirm that the developed foam has good mechanical properties, it was necessary to examine the strut integrity in all the stages of development. Figure 5 shows the morphology and strut integrity of chromized foam. It was observed from the results that the chromized foam did not have any strut disconnections or failures; hence, it was confirmed that the chromizing process did not destroy the strut integrity. The chromized foam had strut thickness of around 250–260 μm and a minimum weight gain of 6.75 g. All the results evaluated for the process capability of the chromizing process showed consistent and promising results. Hence, the chromizing process can be used to deposit the chromium on the sintered Ni foam.

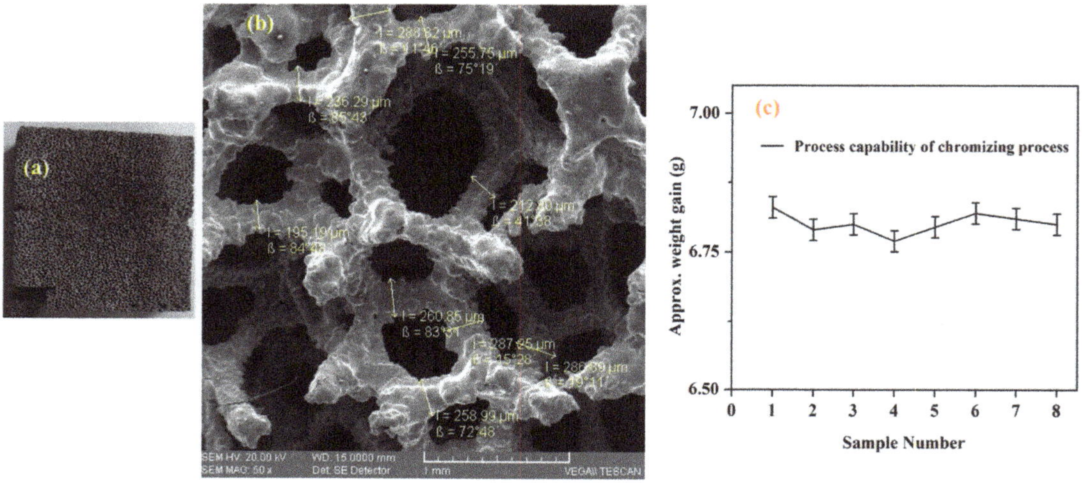

Figure 5. (**a**) Chromized foam and its (**b**) strut thickness and (**c**) process capability.

3.3. EDS and XRD Analysis of Ni-Cr Alloy Foam

To evaluate the bonding and the phase formations between the Ni and Cr elements, EDS and XRD analyses were performed on the foam. Figure 6 shows the line scanning and EDS spectrum of developed Ni-Cr alloy foam. It can be noted from the results that the Ni and Cr elements were uniformly distributed throughout the foam and the weight% of Ni was 71.56, Cr was 22.65 and oxygen was 2.17. This result indicated that, though the bonding occurred between the Ni and Cr elements, oxidation of Cr also happened. This oxidation phenomena can be attributed to an improper inert-gas flow during chromizing and can be controlled by a thorough cleaning of the furnace and sample, and by maintaining the argon gas flow rate at 2 bar. EDS analysis alone is not sufficient to confirm the bonding between the Ni and Cr elements; hence, to confirm the bonding and to estimate the phase formations, XRD analysis was used. The results were compared with reference patterns in X'Pert HighScore and confirmed the formation of Ni_2Cr and $Ni_2Cr_4O_7$ compounds. In addition, to support the results, the XRD spectrum of the developed foam was compared with a Ni-Cr (60:40%) solid sample. It can be observed from the results that major peaks of both spectrums were matching and, hence, it should be concluded that the developed foam has good bonding between Ni and Cr. It can also be noted from the XRD results that the Ni-Cr foam had body-centered cubic structure. Thus, the results confirmed the complete diffusion of face-centered Cr elements into the body-centered Ni foam and the bonding between them. Since the developed foam had a good amount of Cr, it can be used for corrosion-resistance and electrical-resistance applications.

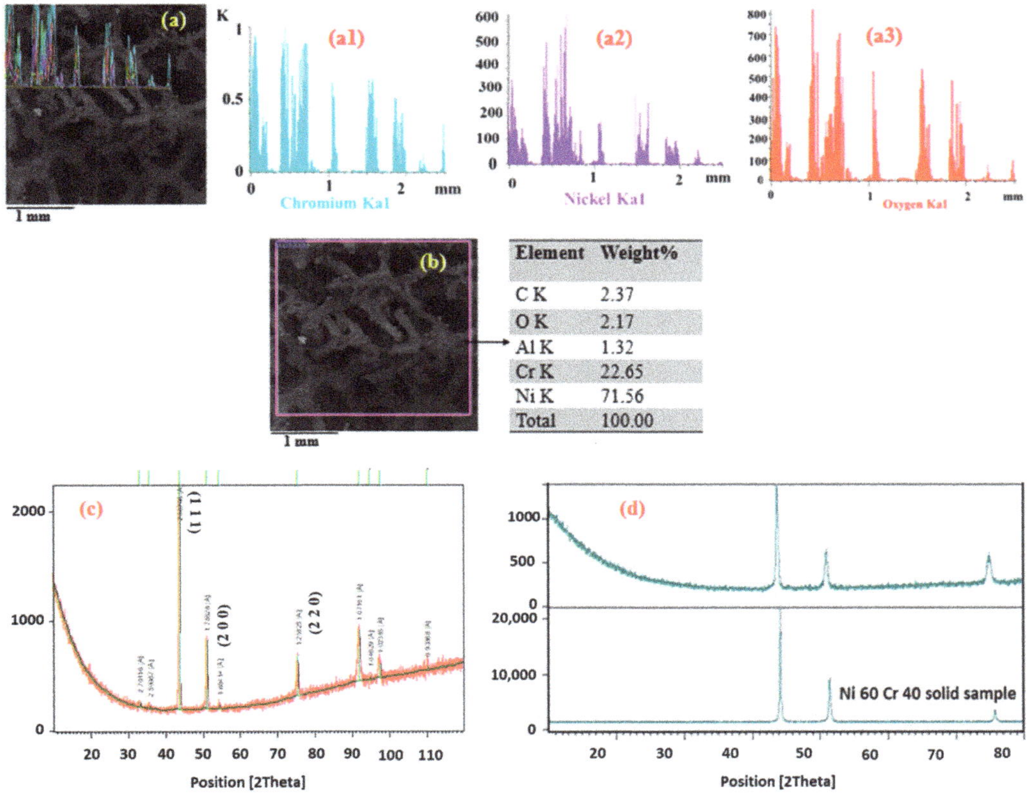

Figure 6. (**a**) Line scanning of Ni-Cr foam along 2.5 mm length; a1-Chromium, a2-Nickel, a3-Oxygen (**b**) elemental mapping; (**c**) XRD spectrum of developed foam; (**d**) comparison of foam with solid sample.

3.4. Compressive Strength of Ni-Cr Alloy Foam

To study the non-linear compressive behavior of the developed foam, a uniaxial compression test was carried out using a UTM. The deformation of the Ni-Cr foam can be described using three stages. The first stage is called the linear elasticity portion, and in this region the deformation behavior follows Hooke's law. The second stage is called the collapse plateau. Here, the plastic deformation and the fracture of cell walls progresses simultaneously until a distinct peak followed by a small stress drop occurs. It can be seen from Figure 7 that the stress generated in the foam is increased up to some plastic deformation stress and then becomes constant. This region of constant stress is called the crush plateau region and it defines the behavior of an ideal energy absorber.

The third stage of deformation is called the densification of the foam. Once the applied stress exceeds the crush plateau region, the foam will start to compress at a constant stress up to 35%–50% of strain. The densification region of the curve acts as a safety backup zone in energy-absorption applications and it allows unexpected energy to be absorbed with an increasing resistance to the impact loads. It is observed from the results that the maximum compressive strength of the developed foam was 10.21 MPa at a maximum displacement of 0.8. Table 2 compares the compressive strength and porosity% of the developed foam with other foams. It can be concluded that the developed Ni-Cr foam has better compressive properties and porosity than the earlier developed foams and thus can have better potential applications than the alternative materials.

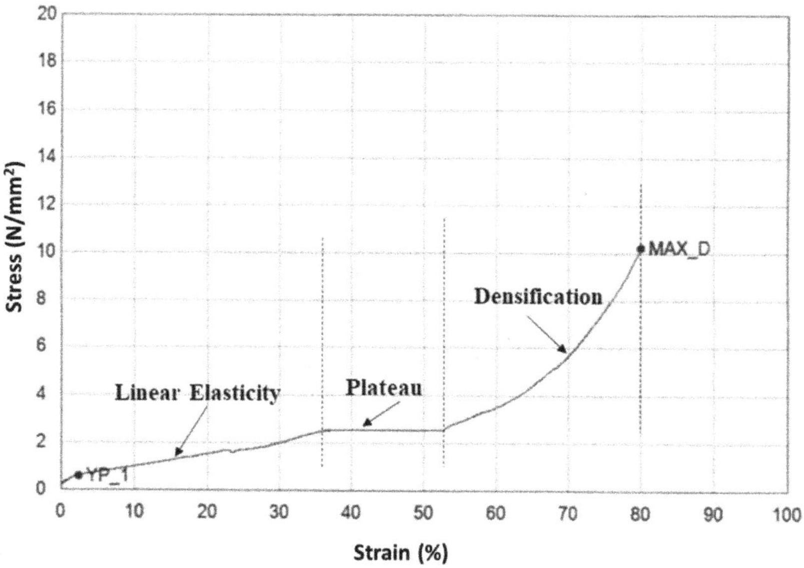

Figure 7. Stress–strain curve of Ni-Cr foam subjected to uniaxial compressive loads.

Table 2. Comparison of compressive strength and porosity with earlier studies.

Type of Foam	Max Compressive Strength	Porosity%	References
Ni-Cr alloy foam	10.21 MPa	91%	Current study
Ni-Cr-Fe alloy foam	3.5 MPa	85%	[15]
AZ91 Mg alloy foam	2.2 MPa	83%	[17]
SS316L open-cell foam	3.0 MPa	80%	[5]
Open-cell copper foam	0.7 MPa	93%	[2]

3.5. Porosity of Foam at Different Processing Stages

In order to examine the effects of the coating techniques on the microstructure of the PU precursor, the porosity% was calculated at different processing stages of the Ni-Cr alloy foam. The theoretical porosity of the developed foam was determined by Equation (1) and the water displacement method. It can be observed from Figure 8 that the porosity% of the foam is decreased from the ELN stage to the UAEPN stage, and then again to the chromizing stage. This change in porosity% can be attributed to an increase in strut thickness from the ELN stage to the chromizing stage. During the sintering process, the bonding between plated Ni particles happens and thus a negligible decrease in the porosity % can be observed.

3.6. Permeability and Pressure Drop of Ni-Cr Alloy Foam

In order to estimate the ease of turbonycoil flow through the developed foam, permeability was calculated at different processing stages of the foam. It can be noted from the results that the permeability decreased from the precursor foam to the UAEPN foam, and then again to the chromized foam. This decrease can be attributed to a decrease in the porosity% with the increase in strut thickness. It is observed from Figure 9 that the foam had good permeability at all the stages of processing; hence, it can be concluded from the study that the developed foam can be used for filtering applications.

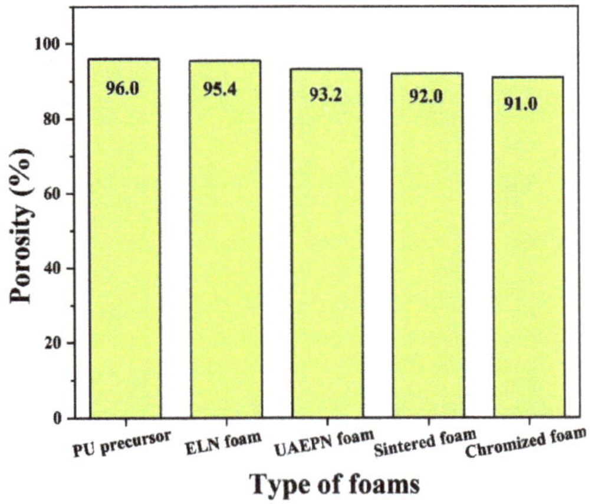

Figure 8. Porosity % of Ni-Cr foam at different processing stages.

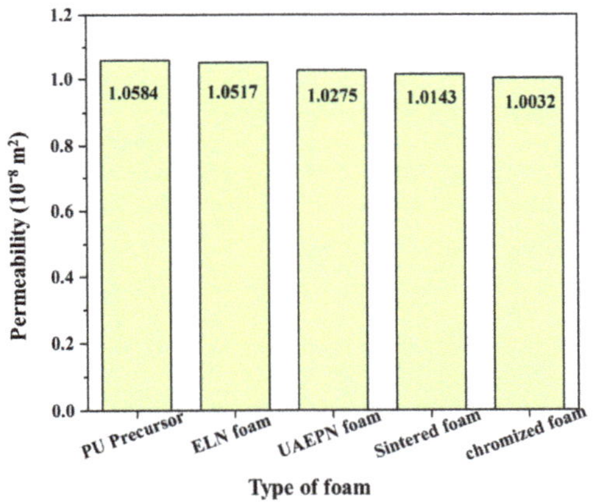

Figure 9. Permeability of Ni-Cr foam at different processing stages.

Though the porosity% and permeability values are satisfactory, the pressure drop through the filtering medium should also be satisfactory to be used in applications as an oil–air separator or filtering media. Hence, the pressure drop through the developed Ni-Cr alloy foam was calculated using Equation (3). For the calculations, turbonycoil dynamic viscosity was considered as 2.544×10^{-5} poise. Turbonycoil was supplied to the foam from a height of 220 mm with a velocity of 0.62 m/s and flow rate of 250 mL/s. It was observed from the results that, for a porosity of 0.91, the pressure drop through the developed Ni-Cr alloy foam was 0.0258%. This pressure drop is negligible; hence, it can be concluded from the study that the developed foam can be used in applications as a filtering medium.

4. Conclusions

The methodology adopted to synthesize a Ni-Cr alloy foam was discussed. Coating techniques such as electroless Ni coating, ultrasonic-assisted electroplating, polyurethane burn-out and sintering, and pack-cementation techniques were discussed. The surface morphology, coating thickness, theoretical porosity, permeability and pressure drop through foam were analyzed. The following are some of the conclusions from the study.

The ELN technique was used to metallize the precursor and resulted in a minimum weight gain of 2 g and a coating thickness of 15–20 µm.

The UAEPN technique ensured the uniform deposition of Ni throughout the ELN-coated foam and showed a minimum weight gain of 42 g and a coating thickness of 40–60 μm.

The chromizing process was evaluated for uniform deposition of Cr and resulted in a minimum weight gain of 6.75 g.

The XRD spectrum confirmed the bonding between the Ni and Cr elements and the phase formation of Ni_2Cr.

The developed Ni-Cr foam had a porosity of 91% and a permeability of 1.0032×10^{-8} m^2.

Author Contributions: Methodology, R.K.P.; Investigation, P.S. Data. curation, S.P.; Writing—original draft, G.A.; Visualization, V.V.; Supervision, S.R.D.; Project administration, C.S.K.; Funding acquisition, F.F. All authors have read and agreed to the published version of the manuscript.

Funding: Filipe Fernandes acknowledges the CEMMPRE (UIDB/00285/2020) and ARISE (LA/P/0112/2020) projects, sponsored by national funds through the FCT—Fundação para a Ciência e a Tecnologia.

Institutional Review Board Statement: Not applicable.

Informed Consent Statement: Not applicable.

Data Availability Statement: Not applicable.

Acknowledgments: The authors would like to thank the Non-Ferrous Materials Technology Development Centre (NFTDC), Hyderabad, for providing all the research facilities. The work is also supported by Maria Zambrano grants from the Spanish government. In addition, the authors are thankful for Grant PID2019-109340RB-I00 funded by MCIN/AEI/10.13039/501100011033. The authors are also thankful for the technical and human support provided by SGIker (UPV/EHU/ERDF, EU).

Conflicts of Interest: The authors declare no conflict of interest.

References

1. Abidi, M.H.; Moiduddin, K.; Siddiquee, A.N.; Mian, S.H.; Mohammed, M.K. Development of Aluminium Metal Foams via Friction Stir Processing by Utilizing $MgCO_3$ Precursor. *Coatings* **2023**, *13*, 162. [CrossRef]
2. Sutygina, A.; Betke, U.; Hasemann, G.; Scheffler, M. Manufacturing of Open-Cell Metal Foams by the Sponge Replication Technique. *IOP Conf. Ser. Mater. Sci. Eng.* **2020**, *882*, 012022. [CrossRef]
3. Pang, Q.; Wu, G.H.; Xiu, Z.Y.; Chen, G.Q.; Sun, D.L. Synthesis and Mechanical Properties of Open-Cell Ni-Fe-Cr Foams. *Mater. Sci. Eng. A* **2012**, *534*, 699–706. [CrossRef]
4. Tseng, C.H.; Li, Z.; Shiue, A.; Chao, Y.H.; Leggett, G. Evaluation of Foamed Nickel Photocatalytic Filters for an Air Cleaner on Removal of Formaldehyde in the Indoor Environment. *Optik* **2021**, *244*, 167550. [CrossRef]
5. Lim, T.Y.; Zhai, W.; Song, X.; Yu, X.; Li, T.; Chua, B.W.; Cui, F. Effect of Slurry Composition on the Microstructure and Mechanical Properties of SS316L Open-Cell Foam. *Mater. Sci. Eng. A* **2020**, *772*, 138798. [CrossRef]
6. Zhang, X.; Zhang, X.; Li, X.; Ma, M.; Zhang, Z.; Ji, X. Slurry Rheological Behaviors and Effects on the Pore Evolution of Fly Ash/Metakaolin-Based Geopolymer Foams in Chemical Foaming System with High Foam Content. *Constr. Build. Mater.* **2023**, *379*, 131259. [CrossRef]
7. Parveez, B.; Jamal, N.A.; Anuar, H.; Ahmad, Y.; Aabid, A.; Baig, M. Microstructure and Mechanical Properties of Metal Foams Fabricated via Melt Foaming and Powder Metallurgy Technique: A Review. *Materials* **2022**, *15*, 5302. [CrossRef]
8. Hu, L.; Li, Y.; Zhou, X.; Yuan, G. Characterization of As-Cast Microstructure of Aluminum Foams by Melt Foaming Method. *Mater. Lett.* **2022**, *308*, 131112. [CrossRef]
9. Firoozbakht, M.; Blond, A.; Zimmermann, G.; Kaya, A.C.; Fleck, C.; Bührig-Polaczek, A. Analyzing the Influence of the Investment Casting Process Parameters on Microstructure and Mechanical Properties of Open-Pore Al–7Si Foams. *J. Mater. Res. Technol.* **2023**, *23*, 2123–2135. [CrossRef]
10. Yuan, G.; Li, Y.; Hu, L.; Fu, W. Preparation of Shaped Aluminum Foam Parts by Investment Casting. *J. Mater. Process. Technol.* **2023**, *314*, 117897. [CrossRef]
11. Costa, J.M.; Almeida Neto, A.F. de Ultrasound-Assisted Electrodeposition and Synthesis of Alloys and Composite Materials: A Review. *Ultrason. Sonochem.* **2020**, *68*, 105193. [CrossRef] [PubMed]
12. Vainoris, M.; Cesiulis, H.; Tsyntsaru, N. Metal Foam Electrode as a Cathode for Copper Electrowinning. *Coatings* **2020**, *10*, 822. [CrossRef]
13. Oriňakov, R.; Gorejová, R.; Králová, Z.O.; Oriňak, A. Surface Modifications of Biodegradable Metallic Foams for Medical Applications. *Coatings* **2020**, *10*, 819. [CrossRef]

14. Duan, D.L.; Li, S.; Ding, X.J.; Jiang, S.L. Preparation of Ni-Cr Alloy Foams by Electrodeposition Technique. *Mater. Sci. Technol.* **2008**, *24*, 461–466. [CrossRef]
15. Pang, Q.; Wu, G.H.; Sun, D.L.; Xiu, Z.Y.; Zhang, Q.; Hu, Z.L. Compressive Property and Energy Absorption Characteristic of 3D Open-Cell Ni-Cr-Fe Alloy Foams under Quasi-Static Conditions. *Trans. Nonferrous Met. Soc. China* **2012**, *22*, s566–s572. [CrossRef]
16. Michailidis, N.; Stergioudi, F.; Omar, H.; Papadopoulos, D.; Tsipas, D.N. Experimental and FEM Analysis of the Material Response of Porous Metals Imposed to Mechanical Loading. *Colloids Surf. A Physicochem. Eng. Asp.* **2011**, *382*, 124–131. [CrossRef]
17. Kapłon, H.; Blawert, C.; Chęcmanowski, J.; Naplocha, K. Development of Open-Porosity Magnesium Foam Produced by Investment Casting. *J. Magnes. Alloy.* **2022**, *10*, 1941–1956. [CrossRef]
18. Zhang, X.; Zhang, X.; Li, X.; Tian, D.; Ma, M.; Wang, T. Optimized Pore Structure and High Permeability of Metakaolin/Fly-Ash-Based Geopolymer Foams from Al– and H_2O_2–Sodium Oleate Foaming Systems. *Ceram. Int.* **2022**, *48*, 18348–18360. [CrossRef]
19. Ji, C.; Huang, H.; Wang, T.; Huang, Q. Recent Advances and Future Trends in Processing Methods and Characterization Technologies of Aluminum Foam Composite Structures: A Review. *J. Manuf. Process.* **2023**, *93*, 116–152. [CrossRef]
20. Wan, T.; Liu, Y.; Zhou, C.; Chen, X.; Li, Y. Fabrication, Properties, and Applications of Open-Cell Aluminum Foams: A Review. *J. Mater. Sci. Technol.* **2021**, *62*, 11–24. [CrossRef]
21. Jain, H.; Mondal, D.P.; Gupta, G.; Kothari, A.; Kumar, R.; Pandey, A.; Shiva, S.; Agarwal, P. Microstructure and High Temperature Compressive Deformation in Lightweight Open Cell Titanium Foam. *Manuf. Lett.* **2021**, *27*, 67–71. [CrossRef]
22. Li, G.; Chen, Z.; Tan, Z.; Tian, R.; Zhao, Y.; Wang, L.; Li, W.; Liu, Y. Investigation on Microstructure and Properties of Electrodeposited Ni-SiC- Composite Coating. *Coatings* **2023**, *13*, 695. [CrossRef]
23. Eugénio, S.; Silva, T.M.; Carmezim, M.J.; Duarte, R.G.; Montemor, M.F. Electrodeposition and Characterization of Nickel-Copper Metallic Foams for Application as Electrodes for Supercapacitors. *J. Appl. Electrochem.* **2014**, *44*, 455–465. [CrossRef]
24. Ma, X.; Jing, Z.; Feng, C.; Qiao, M.; Xu, D. Research and Development Progress of Porous Foam-Based Electrodes in Advanced Electrochemical Energy Storage Devices: A Critical Review. *Renew. Sustain. Energy Rev.* **2023**, *173*, 113111. [CrossRef]
25. Smorygo, O.; Mikutski, V.; Vazhnova, A.; Hancharou, V.; Tikhov, S.; Janagam, V.K.; Gokhale, A.A. Improving Sintering Kinetics and Compositional Homogeneity of Inconel 625 Superalloy Open-Cell Foams Made by Suspension Impregnation Method. *Trans. Nonferrous Met. Soc. China* **2021**, *31*, 2388–2401. [CrossRef]
26. Nan, X.; Wang, F.; Xin, S.; Zhu, X.; Zhou, Q. Effect of Process Parameters on Electrodeposition Process of Co-Mo Alloy Coatings. *Coatings* **2023**, *13*, 665. [CrossRef]
27. Choe, H.; Dunand, D.C. Synthesis, Structure, and Mechanical Properties of Ni-Al and Ni-Cr-Al Superalloy Foams. *Acta Mater.* **2004**, *52*, 1283–1295. [CrossRef]
28. Bekoz, N.; Oktay, E. Mechanical Properties of Low Alloy Steel Foams: Dependency on Porosity and Pore Size. *Mater. Sci. Eng. A* **2013**, *576*, 82–90. [CrossRef]
29. Reda, Y.; Abdel-Karim, R.; Zohdy, K.M.; El-Raghy, S. Electrochemical Behavior of Ni–Cu Foams Fabricated by Dynamic Hydrogen Bubble Template Electrodeposition Used for Energy Applications. *Ain Shams Eng. J.* **2022**, *13*, 101532. [CrossRef]
30. Tadrist, L.; Miscevic, M.; Rahli, O.; Topin, F. About the Use of Fibrous Materials in Compact Heat Exchangers. *Exp. Therm. Fluid Sci.* **2004**, *28*, 193–199. [CrossRef]

Disclaimer/Publisher's Note: The statements, opinions and data contained in all publications are solely those of the individual author(s) and contributor(s) and not of MDPI and/or the editor(s). MDPI and/or the editor(s) disclaim responsibility for any injury to people or property resulting from any ideas, methods, instructions or products referred to in the content.

coatings

MDPI

Article

Comparative Evaluation of Coated Carbide and CBN Inserts Performance in Dry Hard-Turning of AISI 4140 Steel Using Taguchi-Based Grey Relation Analysis

Mustafa Özdemir [1], Mohammad Rafighi [2,*] and Mohammed Al Awadh [3]

1 Department of Machine and Metal Technology, Yozgat Bozok University, Yozgat 66200, Türkiye; mustafaozdemir58@gmail.com
2 Department of Mechanical Engineering, Başkent University, Ankara 06790, Türkiye
3 Department of Industrial Engineering, King Khalid University, Abha 61411, Saudi Arabia; mohalawadh@kku.edu.sa
* Correspondence: mohammad.rafighi@gmail.com

Abstract: Dry hard-turning is a vital manufacturing method for machining hardened steel due to its low cost, high machining efficiency, and green environmental protection. This study aims to analyze the effect of various machining parameters on cutting forces and surface roughness by employing RSM and ANOVA. In addition, multi-objective optimization (Grey Relation Analysis: GRA) is performed to determine the optimum machining parameters. Dry hard-turning tests were carried out on AISI 4140 steel (50 HRC) using coated carbide and CBN inserts with different nose radii. The results show that the cutting force components are greatly affected by the cutting depth and cutting speed for both cutting inserts. As the level of cutting depth and cutting speed rise, the cutting forces also increase. However, the feed rate was the main factor in surface roughness. A low feed rate and high cutting speed lead to good surface quality. According to the results, CBN inserts exhibited better performance compared to carbide inserts in terms of minimum cutting forces and surface roughness. The lowest radial force (Fx = 55.59 N), tangential force (Fy = 15.09 N), cutting force (Fz = 30.49 N), and best surface quality (Ra = 0.28 μm, Rz = 1.8 μm) were obtained using a CBN tool. Finally, based on the GRA, the (V = 120 m/min, f = 0.04 mm/rev, a = 0.06 mm, r = 0.8 mm) have been chosen as optimum machining parameters to minimize all responses simultaneously in the machining of AISI 4140 steel using both carbide and CBN inserts.

Keywords: dry hard-turning; surface roughness; cutting forces; optimization; grey relation analysis

1. Introduction

Many manufacturers aim to produce the maximum amount of product in the minimum amount of time without compromising the quality of the product [1]. In this context, environmentally friendly hard-turning, which provides significant benefits such as flexibility in operation, lower setup and cycle times, lower operating costs, and less power and energy consumption, is a selectable process. Hard-turning is performed on materials having 40 to 65 HRC (Rockwell Hardness) [2]. The grinding process is an alternative to hard turning. However, hard-turning is more popular than the grinding process due to its high stock removal ability and capacity to produce complex geometries. In addition, it has been shown to be an environmentally friendly process since no cutting fluid is used in dry hard-turning [3–5].

Heat-treated steels have a wide range of applications in the aerospace and automotive industries, but the finishing process for such steels is challenging and expensive. Nowadays, the turning of hardened steel is possible thanks to advances in tool material and coating technology. Hardened steel is carbon steel composed of approximately 2.1% carbon. In addition, the hardened steel can be subjected to different heat treatment processes to achieve the desired hardness [6]. Hardened steels are widely used in the aerospace, automotive, and machine tool manufacturing industries due to their excellent properties, such as high wear resistance, corrosion resistance, and durability [7]. It is desired to produce machine parts with high tolerances, high dimensional accuracy, and superior surface quality in the aerospace and automotive industries [8]. The most widely used steel is quenched and tempered AISI 4140 steel, used in the manufacturing and automotive industries. It is widely used in the manufacture of mechanical parts such as shafts, gears, and balls [9]. Cubic Boron Nitride (CBN) is a suitable cutting tool for machining hardened steels, exhibiting high toughness, hardness, and a stable structure at high temperatures [10]. On the other hand, high tool costs and machinability limitations are the problems that stand in the way of the hard-turning process [11]. PCBN (Polycrystalline Cubic Boron Nitride) and diamond inserts have high production costs and limited stocks. Therefore, it is necessary to choose a good solution in terms of performance and quality.

Surface quality is one of the most crucial points of the hard-turning process. Recent studies show that workpiece/cutting tool characteristics affect surface quality. Therefore, it is imperative to investigate the relationship between workpiece/tool properties and surface quality [12]. Another important factor influencing the hard-turning process is tool geometry. Surface quality, tool wear, heat generation, chip formation, and cutting force are greatly affected by tool geometry [13,14]. The most important step following the material and cutting tool is the selection of proper machining parameters. Cutting speed, feed rate, and cutting depth should be determined according to the material and cutting tool. High temperatures and cutting forces are generated during hard turning [15]. Therefore, reducing the cutting force is another critical point in hard turning. Recent studies on turning AISI 4140 steel are presented below, and the results are analyzed comprehensively. So far, no experimental study has been conducted to compare CBN and carbide tools in dry hard-turning in terms of both cutting force and surface roughness. In the current study, the effects of machining parameters, tool types (CBN and carbide), and geometries on the cutting forces and surface roughness are investigated in hard-turning AISI 4140 steel.

Gürbüz and Gönülaçar [9] used carbide inserts in hard turning of AISI 4140 steel under MQL and dry cutting conditions. They reported that Fc and Ra increased with increasing feed rates, and Fc tended to decrease with increasing shear rates. MQL application reduced Fc and Ra overall compared to dry machining, and Ra increased with increasing feed rate. The minimum Ra is recorded as 1.090 μm for the MQL condition.

Schwalm et al. [14] examined the effect of different tool types, machining parameters, and lubrication strategies on the resulting roughness and tool wear. Based on the results, tool type has an influential effect on surface roughness. The type I PVD-coated carbide tool results in a minimum surface roughness of 0.55 μm.

Tool life, cutting forces, and tool wear mechanisms were analyzed for surface roughness in the Nikam et al. [16] study. The machining forces increased as the input variables increased. Besides, increasing the feed rate augmented surface roughness. The minimum surface roughness was measured at 0.518 μm.

Aouici et al. [17] compared different ceramic inserts, considering cutting force and flank wear. The results revealed the dominant effect of cutting depth on the cutting force components. Generally, CC650WG and CC650 ceramic inserts produce lower cutting forces compared to others. The minimum cutting force (Fx) was recorded as 14.07 N for the CC650WG insert.

According to many studies on AISI 4140 steel, the feed rate was the most significant indicator of surface roughness. Increasing the feed rate sharply enhances surface roughness [18–20].

Upadhyay [21] used Grey Relational Analysis to optimize surface roughness, cutting force, and material removal rate (MRR) based on different machining parameters. According to the results of GRA, the optimum machining parameter combination was found to be 0.8 mm cutting depth, 180 m/min cutting speed, and 0.15 mm/rev feed rate.

In addition, the feed rate was determined to be the most influential parameter on the responses, followed by cutting speed. The minimum Rz was recorded as 0.359 μm.

Elbah et al. [22] studied the performances of the machining environments (dry, conventional wet, and MQL technique) in terms of surface roughness and cutting force. It was stated that the cutting force obtained with the MQL machining process improved significantly when compared to other machining processes. The minimum surface roughness was obtained using dry machining (0.21), followed by MQL machining (0.23). However, the mean surface roughness for all combinations of machining parameters in wet and MQL machining is better than in dry machining.

Tiwari et al. [23] investigated the applicability of solid lubricants considering different conditions, namely, dry, flood cooling, grease, and MoS2. The surface quality improved when using solid lubricants such as grease and MoS2.

Sultana et al. [24] reported that the effects of process parameters on roughness, cutting force, specific cutting energy requirements, and material removal rate (MRR) were investigated using LN2-assisted cryogenic cooling. The performance of LN2-assisted cryogenic turning outperformed without coolant machining in terms of roughness, principal force, and specific cutting energy. Generally, the feed rate was considered the most significant factor in the responses. However, according to the Taguchi-linked GRA, the cutting environment was the most dominant factor, followed by the feed rate.

Nicolodi et al. [25] presented the effect of tool wear on machining forces and surface roughness. The findings show that a high feed rate and cutting speed significantly affect both flank and crater tool wear, and the feed rate was the dominant parameter on surface roughness and machining force.

Meddour et al. [26] studied the impact of tool nose radius, cutting depth, cutting speed, and feed rate on surface roughness and cutting forces. A combination of a large nose radius and a low feed rate results in good surface quality. However, the cutting forces (Fx, Fy, and Fz) are significantly impacted by cutting depth. The minimum surface roughness is measured as 0.24 μm and the minimum Fx is recorded as 17.94 N for a 0.8 mm nose radius.

Zahia et al. [27] also reported the same results regarding the impact of feed rate and cutting depth on surface roughness and cutting forces, respectively.

According to most studies [28–31], it has been concluded that the most important factor in surface roughness is feed rate. However, Akkuş [32] stated that the cutting depth is the most important factor in Ra. On the other hand, Bagal et al. [33] stated that both cutting speed and cutting depth were the most effective parameters for surface roughness, while feed rate was the least effective.

By analyzing the literature, the lowest surface roughness (0.1 μm) is obtained with the ceramic tool (V = 180 m/min, f = 0.08 mm/rev, a = 0.2 mm) [34], while the minimum Ra values for CBN tools (0.17 μm) were obtained when using a 0.8 mm corner radius [35]. The least surface roughness (0.2 μm) value for carbide inserts was obtained with a combination of V = 190 m/min, f = 0.05 mm/rev, and a = 0.05 mm machining parameters [36]. In the presented study, the performance of CBN and coated carbide inserts is compared in terms of minimum surface roughness and cutting forces. The lowest surface roughness for a CBN insert is (0.28 μm) and for a carbide insert is (0.35 μm) with a combination of (V = 180 m/min, f = 0.04 mm/rev, a = 0.18 mm, r = 0.8 mm) which are close to the obtained results in the literature survey. In the previous studies, the minimum cutting force for carbide inserts was calculated as 15.33 N (V = 170 m/min, f = 0.08 mm/rev, a = 0.3 mm) [37]. In this study, the cutting force values for carbide and CBN inserts were calculated as 15.09 N and 25.06 N, respectively, which are in good agreement with the literature survey. Generally, high cutting speed and low feed rate lead to minimum surface roughness, while low cutting depth and cutting speed result in minimum cutting force.

2. Materials and Methods

In this study, AISI 4140 alloy steel was used as a material for the machining tests. Appropriate heat treatments have been carried out to achieve a hardness of 50 HRC. The cylindrical material is 40 mm in diameter and 240 mm in length. A thermo-scientific X-ray fluorescence spectrometer was used to determine the chemical composition of the test workpiece. The chemical composition, thermal, and mechanical properties of AISI 4140 steel are given in Table 1.

Table 1. Chemical composition, mechanical, and thermal properties of AISI 4140 steel.

Chemical Composition (wt. %)						
C	Si	Mn	Cr	Mo	Fe	
0.38–0.43	0.15–0.3	0.7–1	0.8–1.1	0.15–0.25	Balance	
Mechanical and thermal properties						
Density	Young's modulus (at 25 °C)	Poisson's ratio (at 25 °C)	Tensile strength	Yield strength	Specific heat (at 25 °C)	Thermal conductivity (at 25 °C)
7850 kg/m^3	198 GPa	0.3	729.5 MPa	379.2 MPa	470 j/kg°C	42.7 W/m°C

Hard-turning experiments were carried out on the GoodWay GS-260Y CNC lathe using variable cutting speeds, feed, depths of cut, and tool nose radii. Since dry cutting conditions are considered an environmentally friendly process, all experiments were performed without the use of cutting fluid [2–5]. Dry cutting conditions are not only an environmentally friendly process but also economically affordable by neglecting the costs of purchasing and disposing of cutting fluids [38]. Hard-turning is usually performed with PCBN or mixed ceramic tools that are even harder than the machined material and can withstand the tribological conditions of the process [39]. Cubic boron nitride (CBN) tool material, which exhibits high toughness, high hardness, and a stable structure at high temperatures, is suitable for hardened steels. It has been stated that the CBN cutting tool performs better than coated carbide tools in terms of tool wear and surface quality [40,41]. In this study, the performance of the coated carbide (CNMG120404-MK5 WKK10S) and CBN (CNMA 433-2 WCB50) inserts was compared in addition to the effect of machining parameters and cutting conditions on the surface roughness and cutting forces. The experimental setup is given in Figure 1.

Taguchi L9 experimental design was used to determine the effect of machining parameters shown in Table 2 on surface roughness and cutting forces. In this regard, nine experiments were performed for each insert instead of 81 to evaluate the results. Response surface methodology was used in MINITAB 20 software to analyze the effect ratios of machining parameters. Besides, the percentage contribution of each machining parameter to the response variables was examined using ANOVA. Finally, regression analysis established mathematical models for each cutting insert.

Table 2. Machining parameters for hard-turning AISI 4140 steel.

Factor	Level 1	Level 2	Level 3
Cutting insert	Carbide	CBN	-
Cutting speed (m/min)	120	150	180
Feed rate (mm/rev)	0.04	0.08	0.12
Cutting depth (mm)	0.06	0.12	0.18
Nose radius (mm)	0.4	0.8	1.2

The 9129AA-type Kistler force dynamometer was used to measure the force signals during the cutting process. Ra and Rz roughness parameters were measured using the Mitutoyo SJ-400 surface roughness tester. Average surface roughness was measured according to the ISO-4288:2011 standard [42]. The arithmetic height means Ra is one of the most widely accepted measurements for surface roughness. It is a good and practical indicator to evaluate the surface quality. Ra is the mean of the absolute height deviation from the mean line along the sampling length. The 10-point roughness parameter Rz was also used in the present study, as it is more sensitive than Ra to the irregular heights or depths of peaks and valleys. According to the international ISO standard, Rz is the difference between the five maximum peaks' mean and the five minimum valleys along the measured length [43]. The raw material's surface roughness was measured at 3.07 μm.

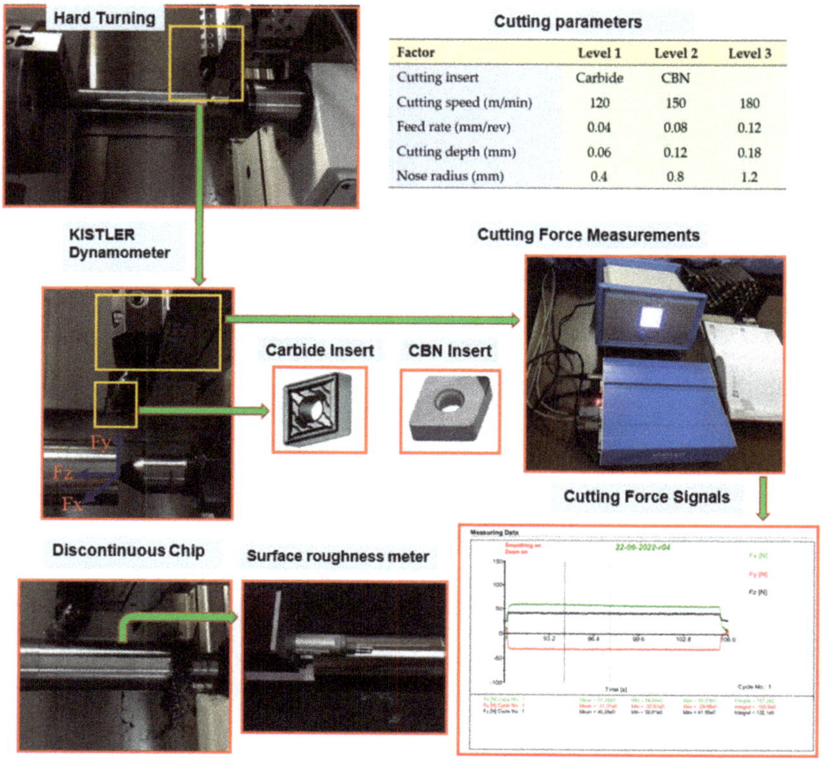

Figure 1. Experimental setup of the turning process.

3. Results and Discussion

The experimental results during dry hard-turning of AISI 4140 steel using carbide and CBN inserts are given in Table 3. This table presents the measured surface roughness parameters (Ra and Rz) and the measured cutting force components (Fx, Fy, and Fz). Here Fx is the radial force, Fy is the tangential force, and Fz is the cutting force.

The raw material's surface roughness was measured at 3.07 μm. However, after the hard-turning process, Ra for carbide inserts is recorded in the range of 0.35 μm and 0.63 μm, while for CBN inserts it is 0.28 μm and 0.59 μm. Rz for carbide inserts varies between 2.4 μm and 4.4 μm, while for CBN inserts it varies between 1.8 μm and 3.8 μm. The cutting force components for the carbide insert are 25.06–198.56 N, while for the CBN insert, they are 15.09–94.36 N. The results showed that the cutting force components are very sensitive to the cutting depth, whereas the surface roughness is greatly affected by feed rate. The feed and cutting depth are functions of the chip cross-sectional area. Thus, as the feed increases, the chip area (undeformed chip thickness) in the machining area augments, which results in a higher cutting force and surface roughness [16].

Tool material is another important factor affecting cutting force. There was a significant reduction in cutting forces using the CBN tool. In terms of radial force (Fx), tangential force (Fy), and cutting force (Fz), the CBN insert presented, respectively, 103%, 62%, and 40% better performance (lower cutting forces) than the carbide inserts. In addition, it was observed that the mean surface roughness for CBN inserts (0.427 μm) is 14% lower than for carbide inserts (0.485 μm). The next important factor affecting the surface roughness is the cutting speed. Cutting speed is one of the essential controlled machining parameters for hard turning [44]. Higher cutting speeds increase the cutting force components; on the contrary, the surface roughness drops as the cutting speed rises.

The minimum cutting force components were determined with a CBN tool (V = 120 m/min, f = 0.04 mm/rev, a = 0.06 mm, r = 0.4 mm). Generally, the radial force is greater than the tangential and cutting forces for both cutting inserts. Nikam et al. [16] also stated that the radial force is the highest force component. The possible reason behind the higher radial force generation was the selection of a lower cutting depth than the insert nose radius [45].

Table 3. Cutting forces and surface roughness results for different machining parameters.

Carbide Insert									
NO	V (m/min)	f (mm/rev)	a (mm)	r (mm)	Fx (N)	Fy (N)	Fz (N)	Ra (μm)	Rz (μm)
1	120	0.04	0.06	0.4	111.36	25.06	43.39	0.44	3.1
2	120	0.08	0.12	0.8	129.79	42.98	55.49	0.52	3.6
3	120	0.12	0.18	1.2	151.88	57.25	71.32	0.63	4.4
4	150	0.04	0.12	1.2	150.52	50.67	63.49	0.39	2.7
5	150	0.08	0.18	0.4	174.02	66.69	80.88	0.49	3.5
6	150	0.12	0.06	0.8	126.89	29.46	50.55	0.58	4.1
7	180	0.04	0.18	0.8	198.56	75.97	91.28	0.35	2.4
8	180	0.08	0.06	1.2	141.63	35.69	58.23	0.45	3.1
9	180	0.12	0.12	0.4	172.39	59.78	72.98	0.52	3.6
CBN Insert									
NO	V (m/min)	f (mm/rev)	a (mm)	r (mm)	Fx (N)	Fy (N)	Fz (N)	Ra (μm)	Rz (μm)
1	120	0.04	0.06	0.4	55.59	15.09	30.49	0.39	2.6
2	120	0.08	0.12	0.8	63.78	25.76	38.69	0.46	3.0
3	120	0.12	0.18	1.2	73.59	34.67	50.28	0.59	3.8
4	150	0.04	0.12	1.2	76.09	32.69	45.68	0.33	2.1
5	150	0.08	0.18	0.4	84.67	37.79	65.19	0.44	2.9
6	150	0.12	0.06	0.8	62.87	19.67	36.66	0.51	3.3
7	180	0.04	0.18	0.8	94.36	48.89	61.59	0.28	1.8
8	180	0.08	0.06	1.2	70.56	23.49	41.49	0.40	2.6
9	180	0.12	0.12	0.4	86.01	35.79	50.89	0.45	2.9

3.1. Main Effects Plot and ANOVA Results

Main effect plots showing the effects of cutting speed, feed rate, cutting depth, and tool nose radius on Fx and Ra are given in Figure 2 according to Taguchi's "the lowest is the best" approach. According to the main effects plot, cutting depth and cutting speed have a dominant effect on the Fx, Fy, and Fz for both cutting inserts. As the value of cutting depth and cutting speed augments, the responses also increase sharply due to the removal of large amounts of material from the workpiece. Furthermore, feed rate has a superior impact on Ra and Rz, followed by cutting speed. Increasing the feed rate sharply enhances the surface roughness. In contrast, increasing the cutting speed reduces the surface roughness. It was concluded that the cutting depth and nose radius have a negligible effect on the Ra.

The primary purpose of the analysis of variance is to show which independent variables significantly affect the dependent variables. The ANOVA results for both cutting inserts are given in Table 4, which shows the contribution effect of each machining parameter on the responses. The machining parameters with a p-value lower than 0.05 are statistically significant. Due to the negligible effect of factor interactions or squares, only the linear resource is used in the ANOVA analysis. According to the ANOVA results for carbide inserts, cutting depth, with 83.75%, 74.71%, and 58.65% contributions, is the most dominant factor on the Fy, Fz, and Fx, respectively. It was followed by cutting speed with 40.10%, 24.50%, and 14.82% contributions on the Fx, Fz, and Fy, respectively. For a CBN insert, the contribution effect of cutting depth on the Fy, Fz, and Fx is 76.47%, 74.32%, and 53.68%, respectively. It was followed by cutting speed with 44.59%, 20.47%, and 18.91% contributions on the Fx, Fy, and Fz, respectively. Increasing the cutting depth and cutting speed causes higher cutting forces due to removing a large amount of material. Therefore, a combination of low cutting depth and cutting speed is preferable to reduce the cutting forces.

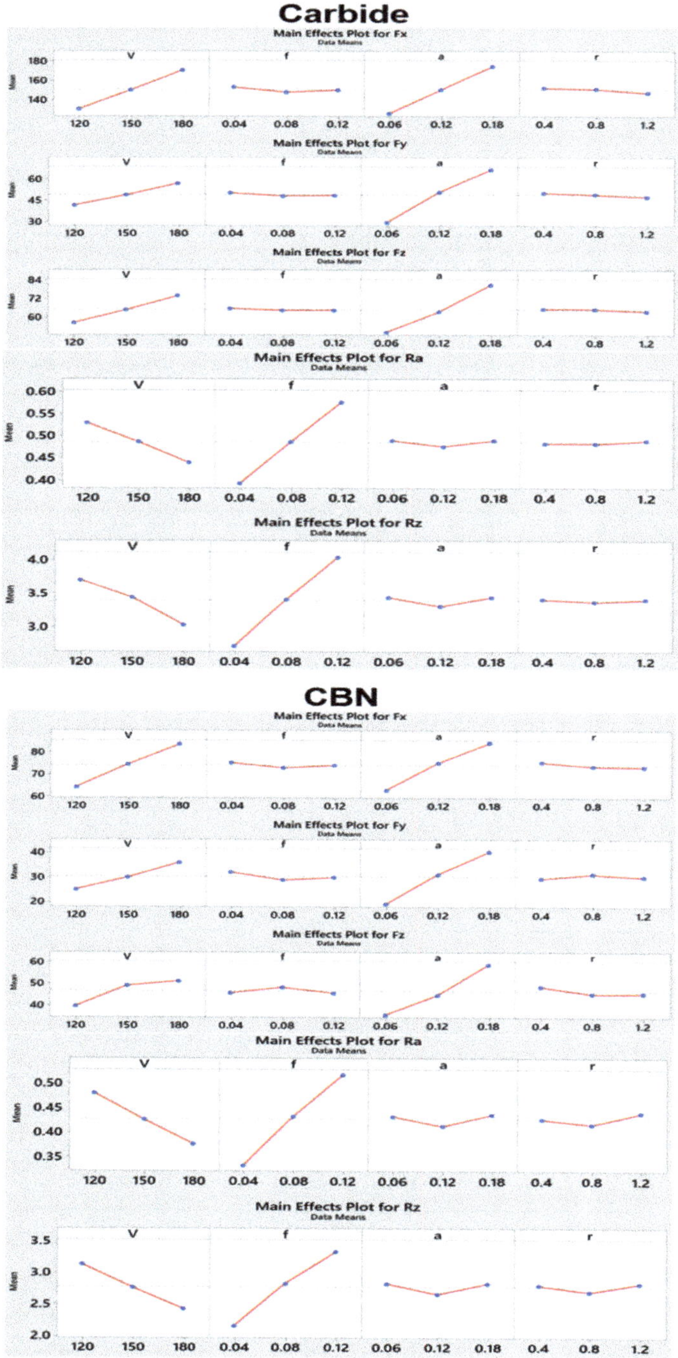

Figure 2. Main effects plot for carbide inserts and CBN inserts.

Table 4. Analysis of variance for cutting forces and surface roughness.

Carbide Insert										
	Fx		Fy		Fz		Ra		Rz	
Source	***p*-Value**	**Cont.**	***p*-Value**	**Cont.**	***p*-Value**	**Cont.**	***p*-Value**	**Cont.**	***p*-Value**	**Cont.**
Linear	0.000	99.52%	0.000	99.19%	0.000	99.46%	0.000	99.38%	0.001	98.55%
V	0.000	40.10%	0.001	14.82%	0.000	24.50%	0.000	19.28%	0.002	20.52%
f	0.229	0.24%	0.388	0.19%	0.440	0.10%	0.000	80.00%	0.000	78.03%
a	0.000	58.65%	0.000	83.75%	0.000	74.71%	1.000	0.00%	1.000	0.00%
r	0.104	0.53%	0.216	0.44%	0.337	0.16%	0.454	0.11%	1.000	0.00%
Error		0.48%		0.81%		0.54%		0.62%		1.45%
Total		100.00%		100.00%		100.00%		100.00%		100.00%
CBN Insert										
	Fx		Fy		Fz		Ra		Rz	
Source	***p*-Value**	**Cont.**	***p*-Value**	**Cont.**	***p*-Value**	**Cont.**	***p*-Value**	**Cont.**	***p*-Value**	**Cont.**
Linear	0.000	98.92%	0.001	97.85%	0.009	94.55%	0.002	97.60%	0.003	96.62%
V	0.000	44.59%	0.003	20.47%	0.020	18.91%	0.003	23.43%	0.005	25.56%
f	0.473	0.17%	0.284	0.82%	0.994	0.00%	0.000	73.76%	0.001	71.00%
a	0.000	53.68%	0.000	76.47%	0.002	74.32%	0.850	0.02%	1.000	0.00%
r	0.252	0.48%	0.701	0.09%	0.381	1.32%	0.465	0.39%	0.806	0.06%
Error		1.08%		2.15%		5.45%		2.40%		3.38%
Total		100.00%		100.00%		100.00%		100.00%		100.00%

The feed rate is the dominant factor in the surface roughness of the carbide insert, with 80% and 78.03% contribution effects for Ra and Rz, respectively. Cutting speed is also a significant indicator of the surface roughness for carbide inserts, with contribution effects of 19.28% and 20.52% for Ra and Rz, respectively. For the CBN insert, the impact of feed rate on the Ra and Rz was 73.76% and 71%, respectively. However, the contribution effect of cutting speed on Ra and Rz is 23.43% and 25.56%, respectively. Increasing the feed rate enlarges the chip area in the turned workpiece, which leads to higher surface roughness. In addition, low cutting speed causes the formation of built-up edges (BUE) on the cutting tool that deteriorate surface quality. Therefore, a combination of a low feed rate and a high cutting speed should be chosen to obtain good surface quality.

3.2. *Surface Graphs*

The 3D surface plot for carbide and CBN inserts considering the interaction of V-a, which is dominant on the cutting force components, is presented in Figure 3, and the 3-D surface plot for surface roughness based on the interaction of V-f is depicted in Figure 4. The cutting force components increase significantly as the cutting depth level is augmented using both cutting inserts. Cutting speed is the next influential factor in the cutting forces. Similarly, as the cutting speed rises, the Fx, Fy, and Fz also enhance. Most of the researchers obtained the same results, indicating the maximum impact of cutting depth on the cutting forces [46,47].

Aouici et al. [17] reported the considerable effect of cutting depth on the cutting force components in the hard-turning of AISI 4140 steel using different ceramic inserts. Tzotzis et al. [47] stated that the dominant effect of cutting depth on the cutting force. İynen et al. [46] also reported the dominant effect of cutting depth on the radial force. Zahia et al. [27] showed the remarkable effect of cutting depth on cutting forces. Increasing the cutting depth sharply enhances the cutting forces.

According to the 3-D surface graph, surface roughness increases using both cutting inserts as the feed rate leads, whereas it decreases as the cutting speed rises. Especially with a combination of low cutting speed and a high feed rate, Ra and Rz increase sharply. High friction occurs between the cutting tool and workpiece at low cutting speeds, reducing surface quality. The minimum surface roughness was obtained using high cutting speeds and a low feed rate. The higher cutting speed increases the cutting zone temperature, which leads to the thermal softening of the material, and consequently, easy chip removal from the workpiece results in good surface quality [16]. One of the most important tasks of mass production is the safe and efficient removal of chips, which is difficult when chips are constantly formed. The chip types are changed from continuous to serrated in hard-turning as the cutting speed increases, which results in significant waviness and roughness of the machined surface. The significant effect of the feed rate on the surface roughness was presented in many studies [48,49].

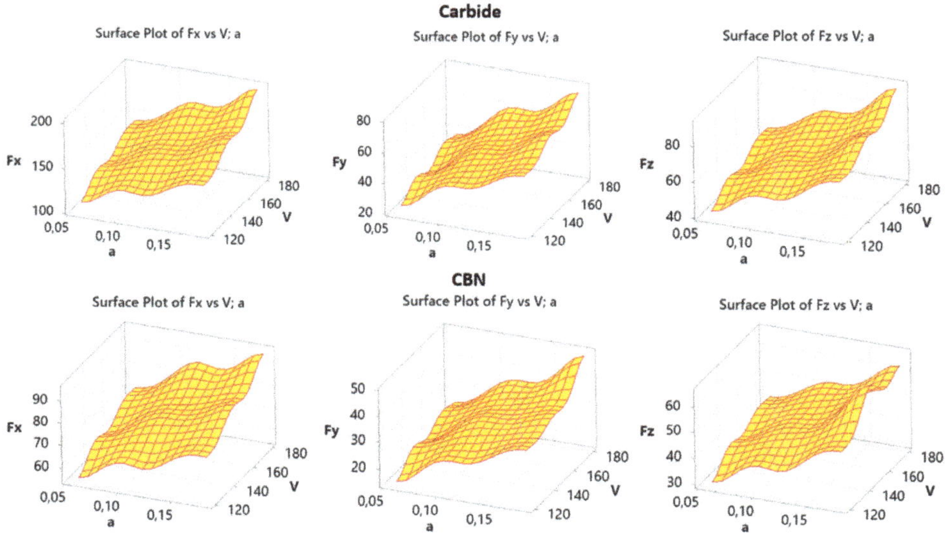

Figure 3. 3-D Surface plot for cutting force components.

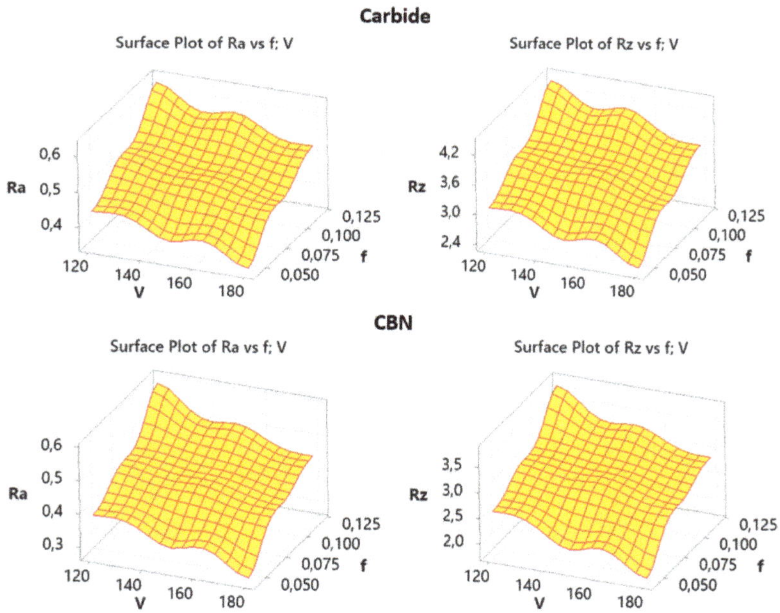

Figure 4. 3-D Surface plot for surface roughness.

İynen et al. [28] stated the dominant effect of feed rate on Ra and Rz. As the feed rate increases, the surface roughness also increases. Mohd et al. [48] reported that the minimum surface roughness was obtained using a low feed rate and high cutting speed. Das [18] and Paengchit [28] also obtained the same results, indicating a dominant effect of feed rate on the surface roughness, followed by cutting speed. Zahia et al. [27] stated the dominant effect of feed rate on surface roughness, followed by cutting speed. They reported that the surface roughness drops as the feed rate decreases and the cutting speed increases. In the presented study, nose radius and cutting depth did not considerably affect the surface quality.

3.3. Prediction Models Using Regression Equations

The multiple regression method was used in this study to develop second-order experimental models at a 95% confidence level for a relationship between machining parameters (V, f, a, r) and output parameters (Fx, Fy, Fz, Ra, Rz) for both cutting inserts. The developed experimental models are given in Equations (1)–(10).

Tool	Regression Equations		R^2	$R^2_{(adj.)}$
Carbide	Fx = 10.64 + 0.6642 V − 38.7 f + 401.6 a − 5.72 r	(1)	99.52%	99.04%
CBN	Fx = 7.86 + 0.3221 V − 14.9 f + 176.7 a − 2.51 r	(2)	98.92%	97.84%
Carbide	Fy = −21.37 + 0.2564 V − 21.7 f + 304.7 a − 3.30 r	(3)	99.19%	98.38%
CBN	Fy = −16.36 + 0.1814 V − 27.2 f + 175.3 a + 0.91 r	(4)	97.85%	95.71%
Carbide	Fz = −6.22 + 0.2905 V − 13.8 f + 253.6 a − 1.75 r	(5)	99.46%	98.93%
CBN	Fz = −1.78 + 0.1917 V + 0.3 f + 190.1 a − 3.80 r	(6)	94.55%	89.09%
Carbide	Ra = 0.5206 − 0.001500 V + 2.292 f + 0.000 a + 0.0083 r	(7)	99.38%	98.77%
CBN	Ra = 0.4861 − 0.001722 V + 2.292 f + 0.028 a + 0.0167 r	(8)	97.60%	95.20%
Carbide	Rz = 3.756 − 0.01111 V + 16.25 f + 0.00 a + 0.000 r	(9)	98.55%	97.09%
CBN	Rz = 3.328 − 0.01167 V + 14.58 f + 0.00 a + 0.042 r	(10)	96.62%	93.24%

Experimental results and predicted results using the abovementioned equations for cutting force components and surface roughness are presented in Table 5.

In addition, a graphical comparison of the experimental and predicted results is depicted in Figure 5 for Fx and Ra using both cutting inserts. Clearly, the presented model fits well with the experimental results since the values are very close.

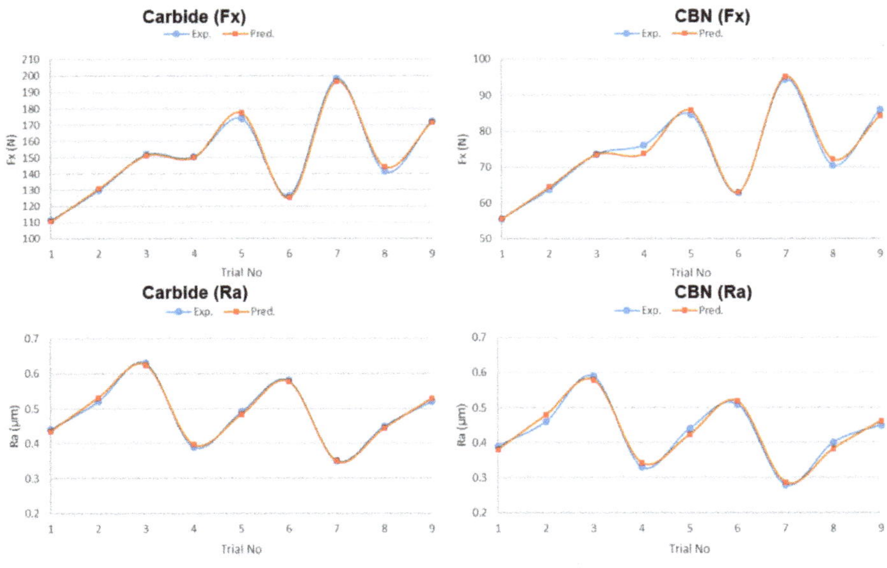

Figure 5. Comparison of experimental and predicted results for Fx and Ra.

Table 5. Experimental and predicted results.

No	Fx (N)		Fy (N)		Fz (N)		Ra (µm)		Rz (µm)	
Carbide Insert										
	Exp.	Pred.	Exp.	Pred.	Exp.	Pred.	Exp.	Pred.	Exp.	Pred.
1	111.36	110.61	25.06	25.49	43.39	42.61	0.44	0.435	3.1	3.08
2	129.79	130.86	42.98	41.59	55.49	56.57	0.52	0.531	3.6	3.73
3	151.88	151.13	57.25	57.68	71.32	70.53	0.63	0.625	4.4	4.38
4	150.52	150.05	50.67	48.82	63.49	65.13	0.39	0.397	2.7	2.74
5	174.02	177.17	66.69	68.88	80.88	81.20	0.49	0.483	3.5	3.39
6	126.89	125.14	29.46	30.13	50.55	49.52	0.58	0.577	4.1	4.04
7	198.56	196.36	75.97	76.12	91.28	89.77	0.35	0.349	2.4	2.41
8	141.63	144.33	35.69	37.37	58.23	58.09	0.45	0.444	3.1	3.06
9	172.39	171.46	59.78	57.42	72.98	74.15	0.52	0.529	3.6	3.71
CBN Insert										
No	Fx (N)		Fy (N)		Fz (N)		Ra (µm)		Rz (µm)	
	Exp.	Pred.	Exp.	Pred.	Exp.	Pred.	Exp.	Pred.	Exp.	Pred.
1	55.59	55.52	15.09	15.20	30.49	31.12	0.39	0.379	2.6	2.53
2	63.78	64.51	25.76	25.00	38.69	41.02	0.46	0.479	3.0	3.13
3	73.59	73.52	34.67	34.79	50.28	50.92	0.59	0.579	3.8	3.73
4	76.09	73.77	32.69	31.89	45.68	45.24	0.33	0.342	2.1	2.21
5	84.67	85.78	37.79	40.59	65.19	59.7	0.44	0.422	2.9	2.76
6	62.87	62.98	19.67	18.83	36.66	35.37	0.51	0.518	3.3	3.36
7	94.36	95.04	48.89	47.48	61.59	63.92	0.28	0.286	1.8	1.84
8	70.56	72.23	23.49	25.72	41.49	39.6	0.40	0.381	2.6	2.44
9	86.01	84.25	35.79	34.43	50.89	54.05	0.45	0.461	2.9	2.99

3.4. Multi-Response Optimization Using Grey Relation Analysis

3.4.1. Grey Relation Analysis Steps

Hard-turning is a multi-response process, and the relationship between the machining parameters and responses is intricate. Therefore, optimum machining parameters based on the multiple response variables can be determined using Grey Relation Analysis (GRA). Besides, the multi-response optimization drops to a single-response optimization using GRA [50,51]. Many researchers used GRA in their studies to optimize multiple responses [52–56]. The following steps should be performed for GRA:

Step 1: Normalization of response parameters

Since response parameter units are different (the cutting force components unit is "N" and the surface roughness unit is "µm"), all responses should be normalized to avoid different units and variability. Thus, all the response values were normalized between 0.00 and 1.00 based on their original values. In this case, the multi-responses can be compared with each other.

In the machining process, responses such as surface roughness and cutting forces are desired to be minimized; hence, the smaller-the-better characteristic is used for normalization considering Equation (11).

$$x_i^*(k) = \frac{\max x_i^0(k) - x_i^0(k)}{\max x_i^0(k) - \min x_i^0(k)} \tag{11}$$

where $i = 1, 2, \ldots, m$, and $k = 1, 2, \ldots, n$. m is the number of experimental data items, and n is the number of output parameters. $x_i^0(k)$ is the original sequence, $x_i^*(k)$ is the sequence after normalization, $\max x_i^0(k)$ is the highest value of $x_i^0(k)$, and $\min x_i^0(k)$ is the lowest value of $x_i^0(k)$.

However, some responses, such as tool life, should be maximized. Therefore, the higher-the-better is used to normalize the value using Equation (12).

$$x_i^*(k) = \frac{x_i^0(k) - \min x_i^0(k)}{\max x_i^0(k) - \min x_i^0(k)} \tag{12}$$

Step 2: Grey relation coefficient

Following the normalization, the grey relational coefficient is calculated to show the relationship between the ideal and actual normalized experimental results. Thus, the grey relational coefficient can be expressed mathematically as Equation (13),

$$\zeta_i(k) = \frac{\Delta_{\min} + \zeta\Delta_{\max}}{\Delta_{0i}(k) + \zeta\Delta_{\max}} \tag{13}$$

where, $\Delta_{0i}(k)$ is the deviation sequence of the reference sequence and the comparability sequence. $\Delta_{0i}(k)$ is calculated using Equation (14).

$$\Delta_{0i}(k) = \|x_0^*(k) - x_i^*(k)\| \tag{14}$$

where, $x_0^*(k)$ shows the reference sequence and $x_i^*(k)$ shows the comparability sequence. $\Delta_{\max}$ and $\Delta_{\min}$ are the maximum and minimum values of the absolute differences $\Delta_{0i}(k)$ of all comparing sequences. ζ is identification or distinguishing coefficient, which is in the range of 0 to 1. However, generally $\zeta = 0.5$ is used.

Step 3: Grey relation grades

According to Equation (15), the grey relational grade is calculated as the average of the grey relational coefficient.

$$\gamma_i = \frac{1}{n}\sum_{k=1}^{n} \zeta_i(k) \tag{15}$$

where, γ_i is the required grey relational grade for i_{th} test and n = number of outputs.

The multi-response optimization problem is converted into a single-response optimization problem employing the grey relational analysis-based Taguchi method [55–57]. The GRG shows the level of correlation between the reference sequence and the comparability sequence and is the total demonstrative of all the quality characteristics. Next, the mean grade relation grade is calculated for each level of the machining parameters. Thus, the optimum machining parameters are determined considering the maximum grey relational grade.

Step 4: ANOVA for GRG values

The Taguchi method could not be used to determine the machining parameters' effect on the multi-responses. Therefore, the ANOVA is performed on the GRG values at a 95% confidence level to determine the most significant factor in the multi-responses.

Step 5: Confirmation test

After determining the optimum machining parameters, it is needed to calculate the grey relation grade to predict and confirm the improvement of the performance factors. The grey relation grade prediction considering the optimum level of machining parameters is calculated using Equation (16).

$$\gamma_{predicted} = \gamma_m + \sum_{i=1}^{k} (\overline{\gamma}_i - \gamma_m) \tag{16}$$

where γ_m is the total mean GRG, $\overline{\gamma}_i$ is the mean GRG at the optimum level of each machining parameter, and k is the number of machining parameters that dominantly affect the multi-responses.

3.4.2. Implementation of Grey Relation Analysis

Firstly, experimental results are normalized for cutting force components and surface roughness using Equation (14). The normalized value for carbide and CBN inserts is given in Table 6.

Table 6. Normalized values for cutting forces and surface roughness.

Carbide Insert					
No	**Fx (N)**	**Fy (N)**	**Fz (N)**	**Ra (μm)**	**Rz (μm)**
1	1.000	1.000	1.000	0.679	0.650
2	0.789	0.648	0.747	0.393	0.400
3	0.535	0.368	0.417	0.000	0.000
4	0.551	0.497	0.580	0.857	0.850
5	0.281	0.182	0.217	0.500	0.450
6	0.822	0.914	0.850	0.179	0.150
7	0.000	0.000	0.000	1.000	1.000
8	0.653	0.791	0.690	0.643	0.650
9	0.300	0.318	0.382	0.393	0.400
CBN Insert					
No	**Fx (N)**	**Fy (N)**	**Fz (N)**	**Ra (μm)**	**Rz (μm)**
1	1.000	1.000	1.000	0.645	0.600
2	0.789	0.684	0.764	0.419	0.400
3	0.536	0.421	0.430	0.000	0.000
4	0.471	0.479	0.562	0.839	0.850
5	0.250	0.328	0.000	0.484	0.450
6	0.812	0.864	0.822	0.258	0.250
7	0.000	0.000	0.104	1.000	1.000
8	0.614	0.751	0.683	0.613	0.600
9	0.215	0.388	0.412	0.452	0.450

Secondly, after computing the deviation sequence, grey relational coefficients are calculated employing Equation (15) using 0.5 for the identification coefficient (ζ) based on the normalized values presented in Table 6. Finally, grey relational grade (GRG) is computed from the results of grey relational coefficients using Equation (15). The grey relation coefficient, grade, and the order for optimum machining parameters considering multi-responses are presented in Table 7.

Based on the ANOVA results, the feed rate is the dominant factor in the surface quality, and the cutting depth is the primary parameter of the cutting force components. The mean grey relation grades are calculated and presented in Table 8 for both cutting inserts. The maximum value of the mean GRG means the minimum cutting force components and surface roughness. As indicated in Table 8, the combination of V1-f1-a1-r2 machining parameters (V = 120 m/min, f = 0.04 mm/rev, a = 0.06 mm, r = 0.8 mm) is selected as the optimum machining parameters in both cutting inserts for the multiple-responses considering the highest mean grey relation grade, which indicates a stronger correlation to the reference sequence and better performance. The difference between the maximum and minimum GRG shows the contribution effect of the machining parameter on the multiple responses; a higher difference means a higher impact. According to Table 8, for carbide and CBN inserts, cutting depth has the dominant effect on the multiple responses, followed by feed rate. The cutting speed and nose radius have a negligible effect on the multi-responses. The main effects plot for the mean grey relation grade is depicted in Figure 6. The red squares show the optimum machining parameter levels for minimizing the multiple responses.

ANOVA results for the GRG value are presented in Table 9. According to the results, cutting depth is the most influential parameter in the multi-responses. It has a 46.93% and 43.36% contribution effect on the carbide and CBN inserts, respectively. It was followed by feed rate, with 40.37% and 40.18% contribution effects, respectively.

Table 7. Grey relation coefficient, grey relation grade, and the order for multi-responses.

Carbide Insert							
	Grey Relation Coefficient					GRG	Order
No	**Fx (N)**	**Fy (N)**	**Fz (N)**	**Ra (μm)**	**Rz (μm)**		
1	1.000	1.000	1.000	0.609	0.588	0.699	1
2	0.703	0.587	0.664	0.452	0.455	0.477	6
3	0.518	0.442	0.462	0.333	0.333	0.348	9
4	0.527	0.498	0.544	0.778	0.769	0.519	2
5	0.410	0.379	0.390	0.500	0.476	0.359	8
6	0.737	0.853	0.770	0.378	0.370	0.518	3
7	0.333	0.333	0.333	1.000	1.000	0.500	5
8	0.590	0.705	0.617	0.583	0.588	0.514	4
9	0.417	0.423	0.447	0.452	0.455	0.366	7
CBN Insert							
	Grey relation coefficient					**GRG**	**Order**
No	**Fx (N)**	**Fy (N)**	**Fz (N)**	**Ra (μm)**	**Rz (μm)**		
1	1.000	1.000	1.000	0.585	0.556	0.690	1
2	0.703	0.613	0.679	0.463	0.455	0.485	6
3	0.519	0.463	0.467	0.333	0.333	0.353	9
4	0.486	0.490	0.533	0.756	0.769	0.506	3
5	0.400	0.427	0.333	0.492	0.476	0.355	8
6	0.727	0.787	0.738	0.403	0.400	0.509	2
7	0.333	0.333	0.358	1.000	1.000	0.504	4
8	0.564	0.668	0.612	0.564	0.556	0.494	5
9	0.389	0.449	0.460	0.477	0.476	0.375	7

Table 8. Mean grey relation grade for carbide and CBN inserts.

Carbide Insert					
Cutting Parameters	Mean GRG			Max and Min Difference	Order
	Level 1	Level 2	Level 3		
V	**0.508**	0.465	0.460	0.048	3
f	**0.573**	0.450	0.411	0.162	2
a	**0.577**	0.454	0.402	0.175	**1**
r	0.475	**0.498**	0.460	0.038	4
	Total Mean GRG: **0.478**				
CBN Insert					
Cutting Parameters	Mean GRG			Max and Min Difference	Order
	Level 1	Level 2	Level 3		
V	**0.509**	0.457	0.458	0.052	3
f	**0.567**	0.445	0.412	0.155	2
a	**0.564**	0.455	0.404	0.160	**1**
r	0.473	**0.499**	0.451	0.048	4
	Total Mean GRG: **0.475**				

Table 9. ANOVA results for carbide and CBN inserts considering GRG values.

Carbide Inserts							
Source	**DF**	**Seq SS**	**Adj SS**	**Adj MS**	**F-Value**	***p*-Value**	**Cont.**
V	1	0.003456	0.003456	0.003456	1.60	0.274	3.54%
f	1	0.039366	0.039366	0.039366	18.28	0.013	**40.37%**
a	1	0.045763	0.045763	0.045763	21.25	0.010	**46.93%**
r	1	0.000308	0.000308	0.000308	0.14	0.724	0.32%
Error	4	0.008615	0.008615	0.002154			8.83%
Total	8	0.097508					100.00%
CBN Insert							
V	1	0.004004	0.004004	0.004004	1.62	0.272	4.50%
f	1	0.035728	0.035728	0.035728	14.46	0.019	**40.18%**
a	1	0.038560	0.038560	0.038560	15.60	0.017	**43.36%**
r	1	0.000748	0.000748	0.000748	0.30	0.611	0.84%
Error	4	0.009886	0.009886	0.002471			11.12%
Total	8	0.088926					100.00%

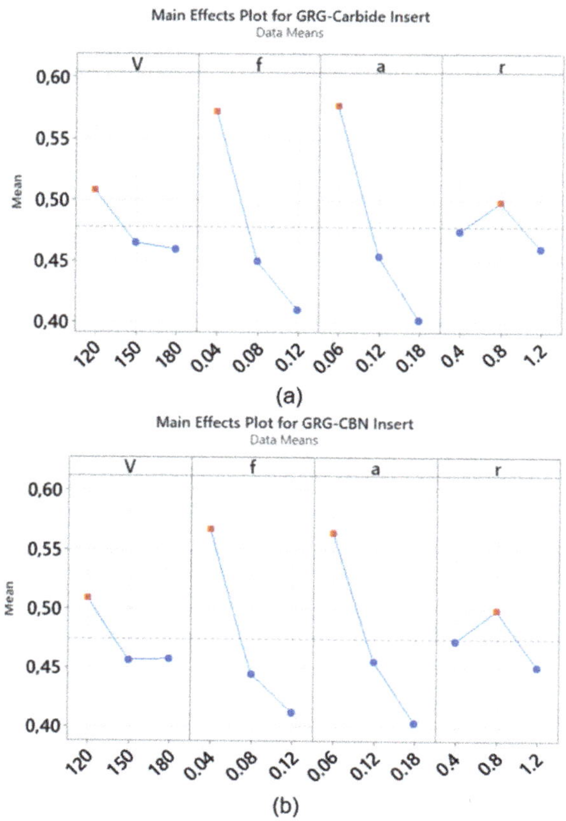

Figure 6. Main effects plots for mean grey relation grade (**a**) carbide insert, (**b**) CBN insert.

According to Equation (16), the predicted grey relation grades are determined using the optimum machining parameters. Table 10 compares the experimental results and the predicted GRG for optimum machining parameters. Based on the results, the cutting force components and surface roughness drop significantly for both cutting inserts using optimum machining parameters. Therefore, considering the initial and optimum machining parameters, an improvement of 0.477 (55.55%) and 0.485 (53.60%) was obtained in GRG for carbide and CBN inserts, respectively. These results are close to those obtained in previous studies [58–60]. Ranganathan and Senthilvelan [53] used the Taguchi-based GRA in the hot turning of stainless steel (type 316), increasing the GRG by 29.7%. Sarıkaya et al. [58] also used Taguchi-based GRA, which improved GRG by 39.4% in turning Haynes 25 alloy. In another study by Sarıkaya et al. [59], Taguchi-based GRA is used in turning AISI 1050 steel, which augments the GRG by 42.9%, considering the optimal machining parameters. Sahoo et al. [60] reported an improvement of 54.5% in GRG from the first machining parameter combination to the optimal machining parameter combination. Pekşen and Kalyon [61] also used Taguchi-based GRA in turning AISI 430 stainless steel, which enhanced the GRG by 69.1%. According to the literature study, multiple-response performance can be significantly increased by using Taguchi-based grey relation analysis in machining hard-to-cut materials.

Table 10. Confirmation test results.

Carbide Insert			
Initial Machining Parameters		**Optimum Machining Parameters**	
		Prediction	**Experiment**
Level	$V_1\ f_2\ a_2\ r_2$	$V_1\ f_1\ a_1\ r_2$	$V_1\ f_1\ a_1\ r_2$
Fx	129.79		100.1
Fy	42.98		19.5
Fz	55.49		33.3
Ra	0.52		0.40
Rz	3.6		2.8
GRG	0.477	0.722	**0.742**
GRG improvement: **0.265** GRG improvement percentage: **55.55%**			
CBN Insert			
Initial machining parameters		**Optimum machining parameters**	
		Prediction	**Experiment**
Level	$V_1\ f_2\ a_2\ r_2$	$V_1\ f_1\ a_1\ r_2$	$V_1\ f_1\ a_1\ r_2$
Fx	63.78		49.7
Fy	25.76		12.2
Fz	38.69		23.9
Ra	0.46		0.33
Rz	3.0		2.2
GRG	0.485	0.714	**0.745**
GRG improvement: **0.260** GRG improvement percentage: **53.60%**			

4. Conclusions

This study investigated the effects of machining parameters and tool geometry on surface roughness and cutting force components in dry hard-turning of AISI 4140 steel. In addition, the performance of coated carbide and CBN inserts was compared. The Taguchi method was used for the design of the experiment; ANOVA was used to determine the contribution effect of independent variables on the dependent variables; and multi-response optimization of the grey relation analysis was utilized to determine the optimum machining parameters for each insert. The findings of the current study are presented below:

- For coated carbide and CBN inserts, Ra was measured in the range of 0.35–0.63 μm and 0.28–0.59 μm, respectively. Besides, for coated carbide and CBN inserts, Rz was measured in the range of 2.4–4.4 μm and 1.8–3.8 μm, respectively. The best surface quality was obtained using a CBN insert (0.28 μm) with a combination of high cutting speed (180 m/min) and low feed rate (0.04 mm/rev). The feed rate is a function of the chip's cross-sectional area. Therefore, by increasing the feed rate, the chip area rises, which results in higher surface roughness. As the cutting speed increases, the built-up edge (BUE) on the cutting tool is reduced, which leads to better surface quality.
- Feed rate was a dominant factor on both Ra and Rz for both cutting inserts. The contribution effect of feed rate on the Ra for carbide and CBN inserts was 80% and 73.76%, respectively. It was followed by cutting speed with 19.28% and 23.43% contribution effects, respectively. Increasing the feed rate value significantly increases the surface roughness components (Ra and Rz). On the contrary, increasing the cutting speed decreases the surface roughness. The undeformed chip thickness increases as the feed rate rises, resulting in high surface roughness. CBN inserts exhibited 14% better performance than carbide inserts in terms of surface quality.

- The cutting force components for carbide and CBN inserts are 25.06–198.56 N and 15.09–94.36 N, respectively. The cutting force components in CBN inserts are lower than in carbide inserts. CBN inserts exhibited 103%, 62%, and 40% better performance than carbide inserts in terms of Fx, Fy, and Fz, respectively. Cutting depth was the dominant factor in all cutting force components. As the cutting depth increases, the Fx, Fy, and Fz augment sharply. Cutting speed is the next important factor in cutting forces. Increasing the cutting speed increases the cutting forces as well. Therefore, a combination of low cutting speed and cutting depth is needed to reduce the cutting forces.
- The developed models can predict the response with very good accuracy, and the graphical comparison of the predicted and experimental results shows perfect agreement between the results.
- The Taguchi-based grey relation analysis was performed to determine the optimum machining parameters. The cutting depth and feed rate have a dominant effect on the multi-responses. The contribution effect of cutting depth on the multi-responses for carbide and CBN inserts is 46.93% and 43.36%, respectively. Besides, the feed rate has a 40.37% and 40.18% contribution effect, respectively. Based on the obtained grey relation grade, a combination of 120 m/min cutting speed, 0.04 mm/rev feed rate, 0.06 mm cutting depth, and 0.8 mm nose radius should be selected for multi-response optimization.

Author Contributions: Conceptualization: M.R. and M.Ö.; Investigation: M.R., M.Ö. and M.A.A.; Experimental Test: M.Ö.; Methodology: M.R. and M.Ö., Software: M.R.; Validation: M.R.; Manuscript draft preparation: M.R., M.Ö. and M.A.A.; Visualization: M.R. and M.Ö.; Paper writing, review, and editing: M.R., M.Ö. and M.A.A. All authors have read and agreed to the published version of the manuscript.

Funding: 1. The author extends their appreciation to the Deanship of Scientific Research, the King Khalid University of Saudi Arabia, for funding this work through the Large Groups Research Project under grant number (RGP.2/163/43). 2. This research was funded by Yozgat Bozok Üniversitesi Bilimsel Araştırmalar Projeleri (BAP), Project No: 6601-FBE/20-352.

Institutional Review Board Statement: Not applicable.

Informed Consent Statement: Not applicable.

Data Availability Statement: Not applicable.

Conflicts of Interest: The authors declare no conflict of interest.

References

1. Awasthi, A.; Saxena, K.K.; Arun, V. Sustainable and smart metal forming manufacturing process. *Mater. Today Proc.* **2021**, *44*, 2069–2079. [CrossRef]
2. Das, A.; Patel, S.K.; Hotta, T.K.; Biswal, B.B. Statistical analysis of different machining characteristics of EN-24 alloy steel during dry hard turning with multilayer coated cermet inserts. *Measurement* **2019**, *134*, 123–141. [CrossRef]
3. Tönshoff, H.; Arendt, C.; Amor, R.B. Cutting of hardened steel. *CIRP Ann.* **2000**, *49*, 547–566. [CrossRef]
4. Klocke, F.; Brinksmeier, E.; Weinert, K. Capability profile of hard cutting and grinding processes. *CIRP Ann.* **2005**, *54*, 22–45. [CrossRef]
5. Chinchanikar, S.; Choudhury, S. Machining of hardened steel—Experimental investigations, performance modeling and cooling techniques: A review. *Int. J. Mach. Tools Manuf.* **2015**, *89*, 95–109. [CrossRef]
6. Chinchanikar, S.; Choudhury, S. Effect of work material hardness and cutting parameters on performance of coated carbide tool when turning hardened steel: An optimization approach. *Measurement* **2013**, *46*, 1572–1584. [CrossRef]
7. Panda, A.; Das, S.R.; Dhupal, D. Surface roughness analysis for economical feasibility study of coated ceramic tool in hard turning operation. *Process Integr. Optim. Sustain.* **2017**, *1*, 237–249. [CrossRef]
8. Das, R.K.; Kumar, R.; Sarkar, G.; Sahoo, S.; Sahoo, A.K.; Mishra, P.C. Comparative machining performance of hardened AISI 4340 Steel under dry and minimum quantity lubrication environments. *Mater. Today Proc.* **2018**, *5*, 24898–24906. [CrossRef]
9. Gürbüz, H.; Emre Gönülaçar, Y. Optimization and evaluation of dry and minimum quantity lubricating methods on machinability of AISI 4140 using Taguchi design and ANOVA. *Proc. Inst. Mech. Eng. Part C J. Mech. Eng. Sci.* **2021**, *235*, 1211–1227. [CrossRef]
10. Chou, Y.K.; Evans, C.J. Tool wear mechanism in continuous cutting of hardened tool steels. *Wear* **1997**, *212*, 59–65. [CrossRef]
11. Rastorguev, D.; Sevastyanov, A. Diagnostics of chip formation and surface quality by parameters of the main drive current in the hard turning. *Mater. Today Proc.* **2019**, *19*, 1845–1851. [CrossRef]

12. He, K.; Gao, M.; Zhao, Z. Soft computing techniques for surface roughness prediction in hard turning: A literature review. *IEEE Access* **2019**, *7*, 89556–89569. [CrossRef]
13. Tzotzis, A.; García-Hernández, C.; Huertas-Talón, J.L.; Kyratsis, P. Influence of the nose radius on the machining forces induced during AISI-4140 hard turning: A CAD-based and 3D FEM approach. *Micromachines* **2020**, *11*, 798. [CrossRef] [PubMed]
14. Schwalm, J.; Gerstenmeyer, M.; Zanger, F.; Schulze, V. Complementary Machining: Effect of tool types on tool wear and surface integrity of AISI 4140. *Procedia CIRP* **2020**, *87*, 89–94. [CrossRef]
15. Dogra, M.; Sharma, V.; Dureja, J. Effect of tool geometry variation on finish turning—A Review. *J. Eng. Sci. Technol. Rev.* **2011**, *4*, 1–13. [CrossRef]
16. Nikam, B.; Khadtare, A.; Pawade, R. Machinability Assesment of AISI 4140 Hardened Steel Using CBN Inserts in Hard Turning. *Int. J. Mod. Manuf. Technol.* **2021**, *13*, 140–148.
17. Aouici, H.; Elbah, M.; Yallese, M.A.; Fnides, B.; Meddour, I.; Benlahmidi, S. Performance comparison of wiper and conventional ceramic inserts in hard turning of AISI 4140 steel: Analysis of machining forces and flank wear. *Int. J. Adv. Manuf. Technol.* **2016**, *87*, 2221–2244. [CrossRef]
18. Das, S.R.; Dhupal, D.; Kumar, A. Experimental investigation into machinability of hardened AISI 4140 steel using TiN coated ceramic tool. *Measurement* **2015**, *62*, 108–126. [CrossRef]
19. Şahinoğlu, A.; Rafighi, M. Investigation of vibration, sound intensity, machine current and surface roughness values of AISI 4140 during machining on the lathe. *Arab. J. Sci. Eng.* **2020**, *45*, 765–778. [CrossRef]
20. Iynen, O.; Şahinoğlu, A.; Özdemir, M.; Yilmaz, V. Investigation of the effect of cutting parameters on the surface roughness value in the machining of AISI 4140 steel with Taquchi method. *J. Inst. Sci. Technol.* **2020**, *10*, 2840–2849. [CrossRef]
21. Upadhyay, V.V. Machining parameters optimization by grey relational analysis of alloy steel AISI 4140. *PalArch's J. Archaeol. Egypt/Egyptol.* **2020**, *17*, 4107–4121.
22. Elbah, M.; Laouici, H.; Benlahmidi, S.; Nouioua, M.; Yallese, M.A. Comparative assessment of machining environments (dry, wet and MQL) in hard turning of AISI 4140 steel with CC6050 tools. *Int. J. Adv. Manuf. Technol.* **2019**, *105*, 2581–2597. [CrossRef]
23. Tiwari, A.; Makhesana, M.A.; Patel, K.M.; Mawandiya, B.K. Experimental investigations on the applicability of solid lubricants in processing of AISI 4140 steel. *Mater. Today Proc.* **2020**, *26*, 2921–2925. [CrossRef]
24. Sultana, M.; Zaman, P.B.; Dhar, N.R. GRA-PCA coupled with Taguchi for optimization of inputs in turning under cryogenic cooling for AISI 4140 steel. *J. Prod. Syst. Manuf. Sci.* **2020**, *1*, 40–62.
25. Nicolodi, J.H.W.; Consalter, L.A.; Durán, O.; Souza, A.J. Effect of wear progression in an 'S'-type mixed ceramic tool on machining forces and surface roughness in the turning of hardened AISI 4140 steel. *Int. J. Mach. Mach. Mater.* **2019**, *21*, 228–243. [CrossRef]
26. Meddour, I.; Yallese, M.A.; Bensouilah, H.; Khellaf, A.; Elbah, M. Prediction of surface roughness and cutting forces using RSM, ANN, and NSGA-II in finish turning of AISI 4140 hardened steel with mixed ceramic tool. *Int. J. Adv. Manuf. Technol.* **2018**, *97*, 1931–1949. [CrossRef]
27. Zahia, H.; Athmane, Y.; Lakhdar, B.; Tarek, M. On the application of response surface methodology for predicting and optimizing surface roughness and cutting forces in hard turning by PVD coated insert. *Int. J. Ind. Eng. Comput.* **2015**, *6*, 267–284. [CrossRef]
28. Paengchit, P.; Saikaew, C. Effects of Cutting Speed and Feed Rate on Surface Roughness in Hard Turning of AISI 4140 with Mixed Ceramic Cutting Tool. *Key Eng. Mater.* **2018**, *79*, 153–158. [CrossRef]
29. Rafighi, M.; Özdemir, M.; Das, A.; Das, S.R. Machinability investigation of cryogenically treated hardened AISI 4140 alloy steel using CBN insert under sustainable finish dry hard turning. *Surf. Rev. Lett.* **2022**, *29*, 2250047. [CrossRef]
30. Sahinoglu, A.; Rafighi, M. Machinability of hardened AISI S1 cold work tool steel using cubic boron nitride. *Sci. Iran.* **2021**, *28*, 2655–2670. [CrossRef]
31. Rafighi, M. Effects of shallow cryogenic treatment on surface characteristics and machinability factors in hard turning of AISI 4140 steel. *Proc. Inst. Mech. Eng. Part E J. Process Mech. Eng.* **2022**, *236*, 2118–2130. [CrossRef]
32. Akkuş, H. Experimental and Statistical Investigation of Surface Roughness in Turning of AISI 4140 Steel. *Sak. Univ. J. Sci.* **2019**, *23*, 775–781. [CrossRef]
33. Bagal, D.K.; Parida, B.; Barua, A.; Jeet, S.; Sahoo, B.B. Multi-parametric optimization in CNC dry turning of chromoly steel using taguchi coupled desirability function analysis and utility concept. *Int. J. Appl. Eng. Res* **2019**, *14*, 21–26.
34. Aouad, R.; Amara, I. Influence of the cutting condition on the wear and the surface roughness in the steel AISI 4140 with mixed ceramic and diamond tool. *J. Eng. Des. Technol.* **2018**, *16*, 828–836. [CrossRef]
35. Şahinoğlu, A.; Ulas, E. An investigation of cutting parameters effect on sound level, surface roughness, and power consumption during machining of hardened AISI 4140. *Mech. Ind.* **2020**, *21*, 523. [CrossRef]
36. Karaaslan, F.; Şahinoğlu, A. Determination of ideal cutting conditions for maximum surface quality and minimum power consumption during hard turning of AISI 4140 steel using TOPSIS method based on fuzzy distance. *Arab. J. Sci. Eng.* **2020**, *45*, 9145–9157. [CrossRef]
37. Rajeev, D.; Dinakaran, D.; Singh, S. Artificial neural network based tool wear estimation on dry hard turning processes of AISI4140 steel using coated carbide tool. *Bull. Pol. Acad. Sci. Tech. Sci.* **2017**, *65*, 553–559. [CrossRef]
38. Sahoo, S. Review on Hard Turning using Finite Element Method. *J. Eng. Innov. Res.* **2019**, *9*, 61–68.
39. Boing, D.; Zilli, L.; Fries, C.E.; Schroeter, R.B. Tool wear rate of the PCBN, mixed ceramic, and coated cemented carbide in the hard turning of the AISI 52100 steel. *Int. J. Adv. Manuf. Technol.* **2019**, *104*, 4697–4704. [CrossRef]

40. Anand, A.; Behera, A.K.; Das, S.R. An overview on economic machining of hardened steels by hard turning and its process variables. *Manuf. Rev.* **2019**, *6*, 4. [CrossRef]
41. Mir, M.J.; Wani, M.F.; Banday, S.; Mushtaq, S.; Khan, J.; Singh, J.; Saleem, S.S. Comparative assessment of coated CBN and multi-layer coated carbide tools on tool wear in hard turning AISID2 steel. In Proceedings of the TRIBOINDIA-2018 an International Conference on Tribology, Maharashtra, India, 13–15 December 2018.
42. Szwajka, K.; Zielińska-Szwajka, J.; Trzepiecinski, T. Experimental study on drilling MDF with tools coated with TiAlN and ZrN. *Materials* **2019**, *12*, 386. [CrossRef] [PubMed]
43. Gadelmawla, E.S.; Koura, M.M.; Maksoud, T.M.; Elewa, I.M.; Soliman, H.H. Roughness parameters. *J. Mater. Process. Technol.* **2002**, *123*, 133–145. [CrossRef]
44. Panda, A.; Das, S.R.; Dhupal, D. Statistical Analysis of Surface Roughness Using RSM in Hard Turning of AISI 4340 Steel with Ceramic Tool. In *Advances in Industrial and Production Engineering*; Springer: Singapore, 2019; pp. 17–26.
45. Sadik, M.I. Wear development and cutting forces on CBN cutting tool in hard part turning of different hardened steels. *Procedia CIRP* **2012**, *1*, 232–237. [CrossRef]
46. Iynen, O.; Şahinoğlu, A.; Özdemir, M.; Yilmaz, V. Optimization of the effect of cutting parameters on the cutting force in the gradual turning process by Taguchi method. *J. Inst. Sci. Technol.* **2020**, *10*, 1909–1918. [CrossRef]
47. Tzotzis, A.; Garcia-Hernandez, C.; Huertas-Talon, J.L.; Kyratsis, P. 3D FE Modelling of Machining Forces during AISI 4140 Hard Turning. *Stroj. Vestn./J. Mech. Eng.* **2020**, *66*, 467–478. [CrossRef]
48. Mohd, A.; Mohd, A.; Adnan, M.H.; Baba, N.B.; Selamat, Z.A.; Rose, A.N.; Mohamed, S.B. Optimization of surface roughness and tool wear on AISI 4140 using coated Ni-YSZ for CNC turning process. *J. Phys. Conf. Ser.* **2020**, *1532*, 012001. [CrossRef]
49. Szwajka, K.; Trzepieciński, T. An examination of the tool life and surface quality during drilling melamine faced chipboard. *Wood Res.* **2017**, *62*, 307–318.
50. Julong, D. Introduction to grey system theory. *J. Grey Syst.* **1989**, *1*, 1–24.
51. Elsayed, E.; Chen, A. Optimal levels of process parameters for products with multiple characteristics. *Int. J. Prod. Res.* **1993**, *31*, 1117–1132. [CrossRef]
52. Pan, L.K.; Wang, C.C.; Wei, S.L.; Sher, H.F. Optimizing multiple quality characteristics via Taguchi method-based Grey analysis. *J. Mater. Process. Technol.* **2007**, *182*, 107–116. [CrossRef]
53. Ranganathan, S.; Senthilvelan, T. Multi-response optimization of machining parameters in hot turning using grey analysis. *Int. J. Adv. Manuf. Technol.* **2011**, *56*, 455–462. [CrossRef]
54. Vinayagamoorthy, R.; Xavior, M.A. Parametric optimization on multi-objective precision turning using grey relational analysis. *Procedia Eng.* **2014**, *97*, 299–307. [CrossRef]
55. Panda, A.; Sahoo, A.; Rout, R. Multi-attribute decision making parametric optimization and modeling in hard turning using ceramic insert through grey relational analysis: A case study. *Decis. Sci. Lett.* **2016**, *5*, 581–592. [CrossRef]
56. Sivaiah, P.; Chakradhar, D. Multi performance characteristics optimization in cryogenic turning of 17-4 PH stainless steel using Taguchi coupled grey relational analysis. *Adv. Mater. Process. Technol.* **2018**, *4*, 431–447. [CrossRef]
57. Senthilkumar, N.; Tamizharasan, T.; Anandakrishnan, V. Experimental investigation and performance analysis of cemented carbide inserts of different geometries using Taguchi based grey relational analysis. *Measurement* **2014**, *58*, 520–536. [CrossRef]
58. Sarıkaya, M.; Güllü, A. Multi-response optimization of minimum quantity lubrication parameters using Taguchi-based grey relational analysis in turning of difficult-to-cut alloy Haynes 25. *J. Clean. Prod.* **2015**, *91*, 347–357. [CrossRef]
59. Sarıkaya, M.; Yılmaz, V.; Dilipak, H. Modeling and multi-response optimization of milling characteristics based on Taguchi and gray relational analysis. *Proc. Inst. Mech. Eng. Part B J. Eng. Manuf.* **2016**, *230*, 1049–1065. [CrossRef]
60. Sahoo, A.K.; Sahoo, B. Performance studies of multilayer hard surface coatings (TiN/TiCN/Al_2O_3/TiN) of indexable carbide inserts in hard machining: Part-II (RSM, grey relational and techno economical approach). *Measurement* **2013**, *46*, 2868–2884. [CrossRef]
61. Pekşen, H.; Kalyon, A. Optimization and measurement of flank wear and surface roughness via Taguchi based grey relational analysis. *Mater. Manuf. Process.* **2021**, *36*, 1865–1874. [CrossRef]

Disclaimer/Publisher's Note: The statements, opinions and data contained in all publications are solely those of the individual author(s) and contributor(s) and not of MDPI and/or the editor(s). MDPI and/or the editor(s) disclaim responsibility for any injury to people or property resulting from any ideas, methods, instructions or products referred to in the content.

coatings

MDPI

Article

Mechanical Properties and Toughening Mechanisms of Promising Zr-Y-Ta-O Composite Ceramics

Xiaoteng Fu [1,2], Fan Zhang [1,2], Wang Zhu [1,2] and Zhipeng Pi [1,2,*]

[1] Key Laboratory of Low Dimensional Materials Application Technology, Xiangtan University, Ministry of Education, Xiangtan 411105, China; 202021001597@smail.xtu.edu.cn (X.F.); zhangfan15@xtu.edu.cn (F.Z.); wzhu@xtu.edu.cn (W.Z.)
[2] School of Materials Science and Engineering, Xiangtan University, Xiangtan 411105, China
* Correspondence: pizhipengmath@163.com

Abstract: ZrO_2-$YO_{1.5}$-$TaO_{2.5}$ (ZYTO) composite ceramics are considered to be a candidate for next-generation thermal barrier coatings (TBCs) due to their excellent thermal stability and low thermal conductivity in high temperatures; however, the mechanical properties and fracture toughness of the ZYTO system may be shortcomings compared with 7-8YSZ: the traditional TBC. In this study, ZYTO composite ceramics were successfully prepared by chemical coprecipitation reaction, and the microstructure of resulting composites was studied as a function of the doping of M-$YTaO_4$. Mechanical properties, including the density, porosity, hardness and Young's modulus, were all determinate; the toughening mechanism was verified by the crack growth behavior of the Vickers indentation test. The results suggest that M-$YTaO_4$ refined the fluorite phase grain and strengthened the grain interface in the composite ceramic. The thermal mismatch between the second phase and matrix produced residual stress in the bulk and affected the crack propagation behavior. With the increase in M-$YTaO_4$ doping, the grain coarsening and ferroelastic domains were observed in the experiments. The ferroelastic domains with orthogonal polarization directions near the crack tip evidenced the ferroelastic toughening mechanism. The competition among these crack behaviors, such as crack deflection, bridging and bifurcation, dominated the actual fracture toughness of the composite. The best toughening formula was determined in the two-phase region, and the highest indentation fracture toughness was about 42 J/m^2, which was very close to 7-8YSZ's 45 $\pm$ 5 J/m^2.

Keywords: mechanical properties; microstructure; toughness; crack propagation

1. Introduction

TBC has been widely used on engine hot-end components in order to improve the efficiency and thrust-to-weight ratio of aero-turbine engines [1]. It is a system of multi-layer materials that are combined to protect engine hot-end components from high-temperature environments [2]. For TBC materials, a lot of research has been published, mainly focusing on thermophysical and mechanical properties. Low thermal conductivity and high-temperature thermostability can improve engine thermal efficiency. Excellent mechanical properties can improve the durability and wear resistance of TBCs [3].

Many researchers have conducted a lot of research in the field of TBC composites. The c-ZrO_2 and $La_2Ce_2O_7$ solid solution with fluorite-type structure powders were mixed and hot-pressed pyrochlore phase LCZ composites [4]. The material has high relative density, small grain size, higher thermal stability, and lower thermal conductivity, with fracture toughness values in the range of 2.13–2.5MPa$\cdot$m$^{1/2}$. In addition, the Vickers hardness of LC40Z composites ranged from 8.68 $\pm$ 0.87 to 10.99 $\pm$ 0.23 GPa, and the fracture toughness ranged from 1.97 $\pm$ 0.15 to 2.4 $\pm$ 0.14 MPa$\cdot$m$^{1/2}$. The denser microstructure had homogeneous grains and reduced porosity [5]. High-entropy rare earth niobates, the $ReNbO_4/Re_3NbO_7$ composite, was prepared via a solid-state reaction. The high-entropy rare earth niobates exhibited excellent phase stability, higher fracture toughness and hardness, with a fracture toughness of 2.71 ± 0.17 MPa·m$^{1/2}$ and hardness of 9.46 ± 0.24 GPa, respectively. The high-entropy niobates exhibited high coefficients of thermal expansion, which were close to 7YSZ [6]. The quasi-binary $GdNbO_4/Gd_3NbO_7$ composites due to residual stress-activated ferroelastic domain switching and the fracture toughness were significantly improved with a toughness value of 2.76 MPa·m$^{1/2}$, which is currently the best for this series of high-entropy rare earth niobates [7]. These composites have low thermal conductivity, excellent mechanical properties, an appropriate coefficient of thermal expansion, and a comprehensive balance of properties in all aspects. The low thermal conductivity of the material, with a comprehensive balance of properties, provides an option for the design and manufacture of TBC materials and has excellent application potential.

The ceramic of the ZYTO system is a popular candidate for the next generation of thermal barrier coatings, with advantages such as thermodynamic stability [8,9] at high temperatures, resistance to harmful phase transformation, low thermal conductivity [10,11], and potential strength and toughness [12,13]. Therefore, the toughening mechanism of the ZYTO system has aroused widespread concern among scholars. As we know, ferroelastic toughening in the ZYTO system maximizes the toughness of ceramics and enhances their practicability in practical applications [14–17]. However, the fracture toughness of the ZYTO system is higher than some traditional ceramics but not compared with YSZ [18]; this weakness shortens its service life as a thermal barrier coating. Traditional 7-8YSZ's fracture toughness is about 4MPa·m$^{1/2}$ [19]. Recently, it was found that 7YSZ could reach a higher fracture toughness via a finely crystalline technique under sintering conditions at 1600 °C, the microstructure of which exhibits a composite phase composed of m-ZrO_2 and t-ZrO_2 [20,21]. In this paper, we focused on the mechanism that the second phase M-$YTaO_4$, affects toughening by changing the crack behavior; based on this, we attempted to determine the best toughening formula of M-$YTaO_4$ doping.

This paper studies the mechanical properties characterization and crack growth mode of the fluorite field and M-$YTaO_4$ field in the ZYTO system. The M-$YTaO_4$ solid solution was studied by XRD and EDS, the distribution of the second phase was observed under SEM, and several crack propagation behaviors were studied. The fine grain, phase stability, interface binding and residual stress of composites were shown to be influenced by the second-phase doping content. We tried to find some answers to several remaining problems. First, are the crack propagation behaviors and the fracture toughness sensitively dependent on the composition? Second are ferroelastic toughening and second phase toughening in cooperation or competition. Moreover, does the best toughening formula exist for this composite ceramic?

2. Material and Methods

ZYTO composite ceramic powders were prepared by chemical co-precipitation and calcination. First, five groups of YT1–YT5 samples were set in x mol%Ta_2O_5 (x = 10/20/30/40/50) increments. All samples were prepared by a chemical to coprecipitate using precursor solutions of $ZrO(NO_3)_2$ (≥99.99%), $Y(NO_3)_3{\cdot}6H_2O$ (≥99.99%) and $TaCl_5$ (≥99.99%). The pH of the solution was maintained above 10 to ensure the precipitation of all the mixed cations at the molecular level. The corresponding chemical equation for the development is as follows:

$$2ZrO(NO_3)_2 + 3Y(NO_3)_3{\cdot}6H_2O + TaCl_5 + 18NH_3{\cdot}H_2O \rightarrow YTaO_4 + Zr_2Y_2O_7(\text{fluorite}) + 5NH_4CI + 13NH_4NO_3 + 27H_2O$$

We separated the solution and hydroxide precipitate by centrifuge before cleaning twice with ethanol and drying in a drying oven for 12 h. To ensure the full conversion of tantalum-rich hydroxide to oxide, the pyrolysis was then conducted at 1300 °C for 5 h. The obtained powder was pressed into a single shaft at ~200 MPa and sintered in the air at 1500 °C for 10 h. The preparation process is shown in Figure 1.

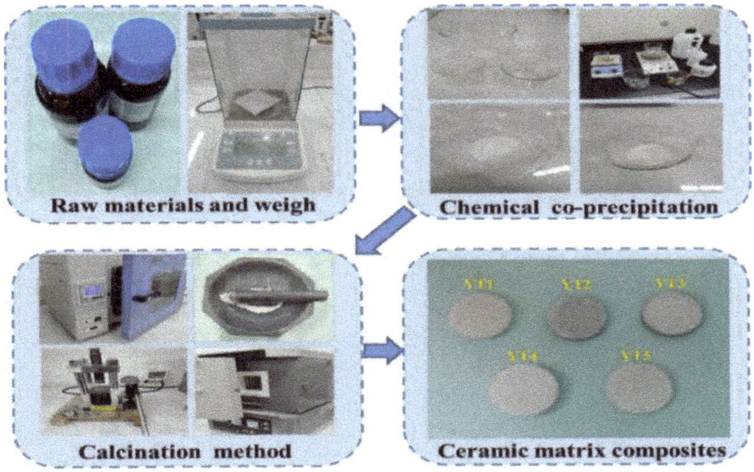

Figure 1. Flow chart of preparation process.

Vickers hardness (H_V) was measured by the indentation method. The indentation hardness test and analysis system (ZHVST-30F, Zhongyan, Shanghai, China) were used for measurement on the polished surface of the sample with a load of 10 s at 29.8 N. At least 6 effective indentations were made per sample. Fracture toughness (K_{IC}) was calculated by Ref. [22] based on the radial crack pattern of Vickers indentation:

$$K_{IC} = 0.16 H_V a^2 c^{-\frac{3}{2}} \tag{1}$$

where H_V is Vickers hardness, a is the half length of the diagonal of the indent, and c is the half crack length measured from the middle of the dent to the crack tip, as shown in Figure 2. Figure 2 shows the secondary electron (SE, TESCAN MIRA3 LMH, Brno, Czech Republic) image, where the diamond indentation shape and cracks can be clearly seen.

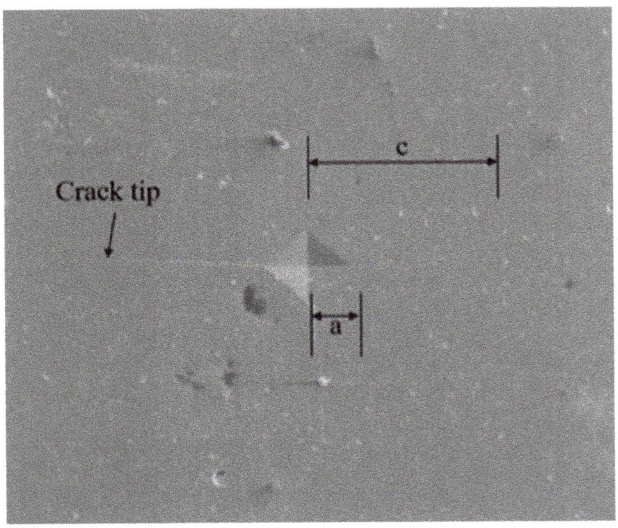

Figure 2. Secondary electron diagram of composite ceramic materials.

The critical energy release rate Γ (fracture energy) refers to the ability to resist fracture, which is the actual crack propagation parameter, as a dynamic evolutionary variable indicating the energy required to propagate a certain length of the crack. The Γ is N·m/m^2 or J/m^2. Therefore, Γ is also understood as the energy provided by the system with each unit area of the crack propagation or the force provided by the system with each unit length of crack propagation. The Γ is also called the driving force or the fracture energy. The formula is:

$$\Gamma = 2\xi^2 P \frac{a^2}{c^3} \quad (2)$$

where the value of ξ is 0.016 (±0.004) calibration constant, P is the loading force, a is the half length of the diagonal of the indent, and c is the average crack length measured from the indentation center.

Young's modulus (E) can be measured by the UMS Advanced Ultrasonic Material Characterization System (UMS-100, Rohde Schwarz, Munich, Germany). Young's modulus can be obtained by the following formula [23,24]:

$$E = \frac{V_L^2 \rho (1+\nu)(1-2\nu)}{1-\nu} \quad (3)$$

$$\nu = \frac{1 - 2\left(\frac{V_T}{V_L}\right)^2}{2 - 2\left(\frac{V_T}{V_L}\right)^2} \quad (4)$$

where V_T is the transverse sound speed, V_L is longitudinal sound speeds, ν is Poisson's ratio and ρ is the density measured by Archimedes' principle, as shown in Table 1.

Table 1. Density of sample.

Sample	Density ρ (g·cm^{-3})
YT1	5.09
YT2	5.93
YT3	6.22
YT4	6.68
YT5	6.95

3. Results and Discussion

3.1. Microstructure Characterization

The X-ray Diffraction (XRD, Ultimate IV, RIGAKU, Tokyo, Japan) patterns of ZYTO composites are shown in Figure 3. The XRD reflections of the fluorite phase and M-$YTaO_4$ are fluorites and scheelite structures. Figure 2 shows that with the increase in M-$YTaO_4$ content, the scheelite peak increases, indicating an increase in the scheelite phase amount. Only the fluorite phase and the M-$YTaO_4$ phase were detected in the composite, revealing no chemical reaction between the fluorite phase and M-$YTaO_4$. We analyzed the phase distribution by means of scanning electron microscopy (SEM, TESCAN MIRA3 LMH, Brno, Czech Republic) equipped with an energy dispersion spectrometer (EDS, X MAX20, Oxford Instruments, Oxford, UK). Surface SEM images of hot-etched ZYTO pellets are shown in Figure 4. In the image of YT1-YT4, grains with light are the M-$YTaO_4$, dark is the fluorite phase, and clear contrasts are observed. With the increase in the doping content, fluorite grain sizes further decreased.

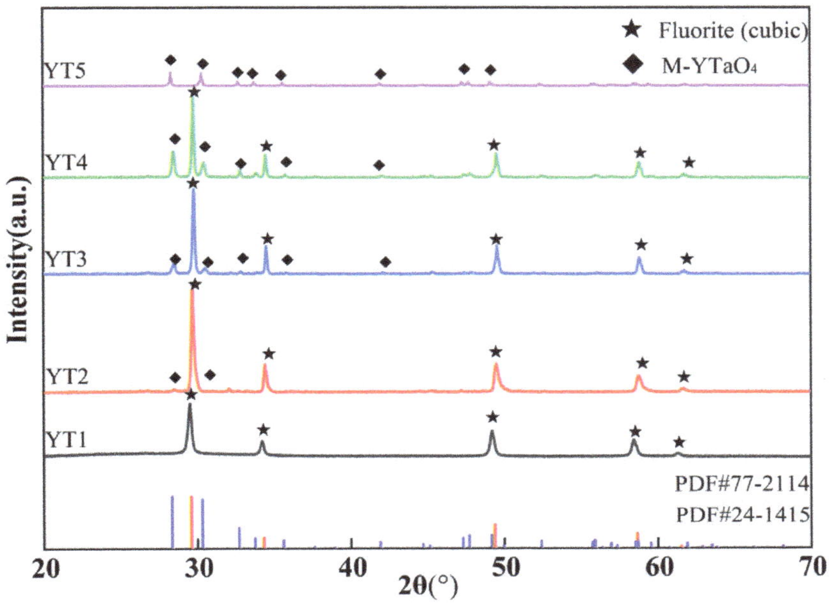

Figure 3. XRD pattern of fluorite phase and M-$YTaO_4$ phase composite at 1500 °C.

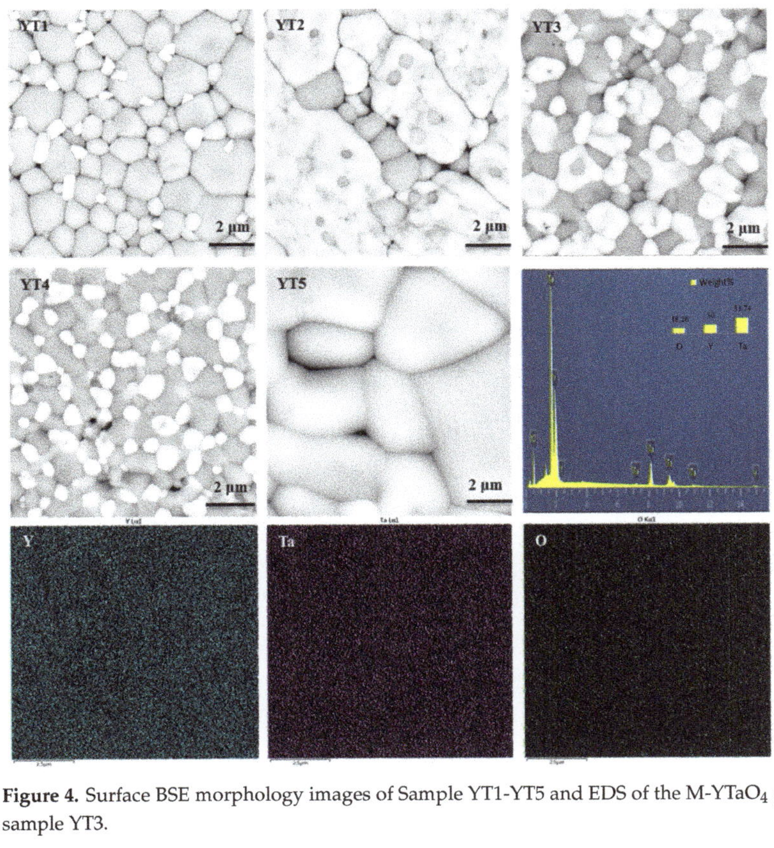

Figure 4. Surface BSE morphology images of Sample YT1-YT5 and EDS of the M-$YTaO_4$ grains in sample YT3.

The linear intercept method [25] was used to estimate the average grain size in the composite. As shown in Figure 5, the average size of fluorite grains showed a decreasing trend with an increase in the second phase. This indicated that the doping of M-$YTaO_4$ could inhibit the growth of fluorite grains during the sintering process. As shown in Figure 3, the interconnection of M-$YTaO_4$ grains occurred in YT3 and YT4, and the M-$YTaO_4$ grains in the composites became gradually larger. It can be noted in particular that when the doping amounts were YT3 and YT4, the fluorite and M-$YTaO_4$ grains in the composites were close in size. According to percolation theory, there is a critical range of the 14~16 vol% second phase volume fraction in the two-phase composites beyond which the second phase interconnection effect occurs [26,27]. In the present study, the M-$YTaO_4$ content was significantly higher than this range, which is why this phenomenon occurred. Furthermore, other studies have shown [28,29] that doping the second phase can refine the grain size of the matrix phase and enhance the mechanical properties of the composites.

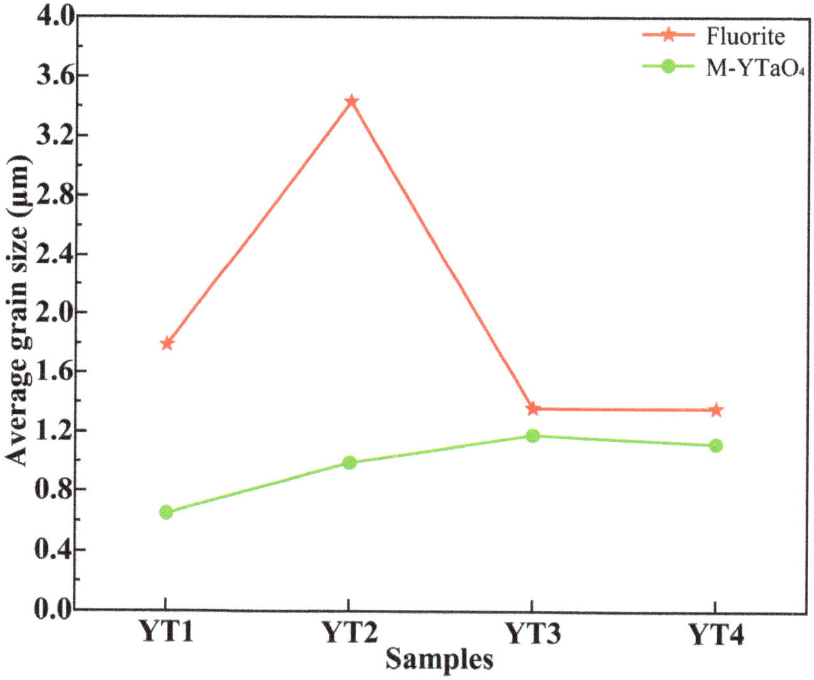

Figure 5. Comparison of the average size of M-$YTaO_4$ and fluorite grains in the composite.

3.2. Hardness and Fracture Toughness

Mechanical properties values obtained by the UMS Advanced Ultrasonic Material Characterization System and the indentation hardness test and analysis system are shown in Table 2. The Vickers hardness of ZYTO composites as a function of M-$YTaO_4$ doping is shown in Figure 6a. Obviously, the hardness decreased almost linearly with the doping of the second phase. The hardness decreased from 10.6 GPa to 4.5 GPa, which was higher than that of 8YSZ without additives (from 1.52 to 2.05 GPa) and was considerably lower than that of 8YSZ with additives CuO-TiO_2 (17.2 to 17.96 GPa) [30].

Table 2. Measured mechanical properties values of samples YT1–YT4.

Sample	$V_L/m{\cdot}s^{-1}$	$V_T/m{\cdot}s^{-1}$	*E*/GPa	*B*/GPa	*G*/GPa	υ	H_V/GPa
YT1	6095.24	3413.33	150.81	110.03	59.3027	0.2716	5.10
YT2	6175.44	3450.98	179.80	131.98	70.6219	0.2730	8.31
YT3	5803.92	3067.36	152.89	131.49	58.5221	0.3062	7.87
YT4	5333.33	2580.65	119.86	130.69	44.4872	0.3471	5.83

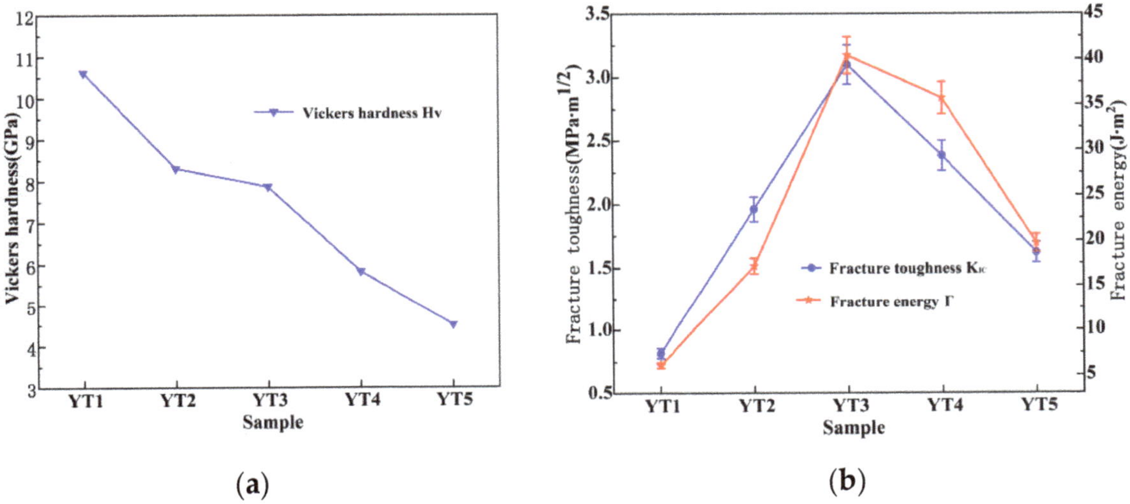

Figure 6. The relationship of Vickers hardness (**a**), fracture toughness and fracture energy (**b**).

Furthermore, due to the low density of the sample and more pores under the same load, the surface of the material was more prone to deformation, resulting in a sharp decline in hardness. Secondly, the porosity and cracks in the sample increased, and the brittleness of the material increased, which also affected the mechanical properties of the material. In addition, pores were apparent in ceramics, which could decrease the hardness of sintered specimens. As shown in Figure 3, a small amount of M-$YTaO_4$ doping was able to enhance the fluorite phase grain boundaries. The fluorite phase ceramic material improved its resistance to deformation in local areas. As a result, the hardness of YT1 was higher. As the amount of M-$YTaO_4$ doping increased, the hardness of the composites YT2-YT4 decreased instead compared to their fluorite phase counterparts. It could, thus, be inferred from the mixing law that the overall hardness approached that of M-$YTaO_4$ phase ceramics as the M-$YTaO_4$ doping content increased.

The trends in fracture toughness and fracture energy calculated by Equations (1) and (2) are shown in Figure 6b. The fracture toughness increased when increasing the M-$YTaO_4$ content and reached the highest value (approximately 3.1 MPa·$m^{1/2}$) with the samples YT3, which was almost 300% higher than that of the fluorite ceramic. At this moment, the fracture energy also reached a peak, and the indentation toughness was approximately equal to 42 J/m^2, which was comparable to 7-8YSZ's 45 $\pm$ 5 J/m^2 [31]. Fracture toughness and fracture energy followed a similar trend with increased doping, both increasing and then decreasing. The original increase in fracture toughness could be attributed to the grain size effect [32].

3.3. Mechanism of Toughening

Based on the above analytical results, we can conclude that the introduction of the second phase M-$YTaO_4$ into fluorite phase ceramics was beneficial in improving the hardness and fracture toughness, although this change was not linearly proportional to the doping concentration. Further analysis of the toughening mechanism of composite ceramics revealed that the increase in fracture toughness was strongly related to residual stress, the interface state, crack propagation mechanism and ferroelastic switching.

3.3.1. Effects of the Residual Stress

In ZYTO composite ceramics, some studies have shown that the CTE (thermal expansion coefficient) of the second phase was higher than that of the matrix. When cooled from sintering temperature to room temperature, the base phase and the second phase produced compressive and tensile stresses, respectively, which is an average (effective) stress level. The displacement of the diffraction peak in the XRD pattern could reflect residual stress. Therefore, the peak displacement can exhibit whether compressive stress or tensile stress is generated. As shown in Figure 7a, the diffraction peaks of the second phase and the matrix moved in an opposite trend with the increase in the doping amount. For example, the M-$YTaO_4$ peaks moved at low angles to produce tensile stresses, while the matrix fluorite phase peaks moved at high angles to produce compressive stresses. This confirmed the formation of tensile stresses in the second phase and compressive stresses in the substrate. The actual situation at different grains and interfaces may be completely different due to the grain size, grain shape in terms of sharp notches, temperature gradient during cooling at the corresponding location, presence of stress micro-concentrators in the form of voids, micro-cracks, etc.

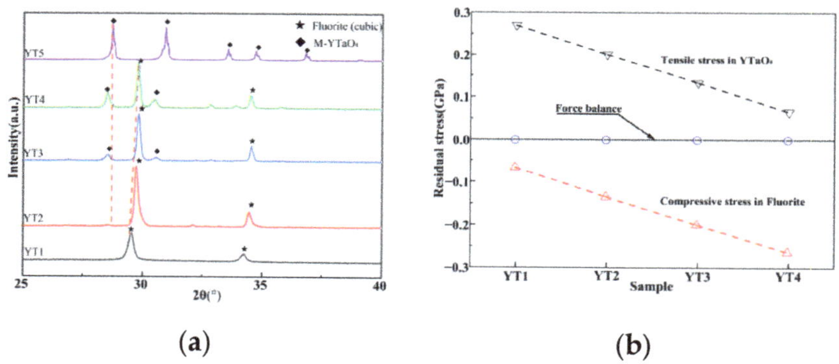

Figure 7. Slow-scan XRD pattern (**a**) of the composite at 2θ = 25°–40° as well as residual stress (**b**).

According to the model proposed by Taya [33], the residual stress in the composite can be calculated by the following equation:

$$\frac{\sigma_s}{E_m} = \frac{-2(1-x)\beta(\alpha_s-\alpha_m)(T_0-T)}{(1-x)(\beta+2)(1+v_m)+3\beta x(1-v_m)} \tag{5}$$

$$\frac{\sigma_m}{E_m} = \frac{2x\beta(\alpha_s-\alpha_m)(T_0-T)}{(1-x)(\beta+2)(1+v_m)+3\beta x(1-v_m)} \tag{6}$$

$$\beta = \frac{1+v_m}{1-2v_s}\frac{E_S}{E_m} \tag{7}$$

where the subscripts s and m represent the second phase and the matrix, respectively, x is the content of the second phase with Young's modulus E, Poisson's ratio ν, thermal expansion coefficient α, tensile stresses σs and compressive stresses σm. T_0 and T represent the room temperature and sintering temperature, respectively. Table 3 lists the values of these parameters for calculating residual stress. The TEC of the fluorite phase is (9.6×10^{-6} K^{-1}), which is lower than that of the M-$YTaO_4$ phase (10.7×10^{-6} K^{-1}) [34]. Young's modulus and Poisson's ratio in the M-$YTaO_4$ phase were (177 GPa and 0.35), and that of the fluorite phase was (210 GPa and 0.3) [35]. Therefore, the calculated residual stresses are shown in Figure 7b. It is obvious that in the composite, the second phase grain was subjected to tensile stress, while the matrix was under compressive stress. With the increase in the second phase content, the tensile stress of the second phase grain decreased, while the compressive stress in the matrix increased. As we know, residual stresses during cooling can heal cracks and improve toughness [36].

Table 3. Calculation parameters of residual stress.

Phase	E (GPa)	v	A (K^{-1})
Fluorite	210	0.30	9.6×10^{-6}
M-$YTaO_4$	177	0.35	10.7×10^{-6}

The synergistic influence of residual stress on crack growth behavior is schematically plotted in Figure 8. When the tensile stress was perpendicular to the direction of the crack extension, and the compressive stress was opposite to the direction of the tensile stress, the crack passed through the second phase, which consumed fracture energy and improved the toughness of the material (mode I). When the tensile stress direction was parallel to the interface, the crack was likely to propagate in the interface plane rather than along the initial path, resulting in crack deflection (Mode II). If the tensile stress was perpendicular to the interface, a new crack originating at the interface or crack bridging (mode III) may have occurred. For composite ceramics with different M-$YTaO_4$ content, their second phase distribution and residual stress distribution are different.

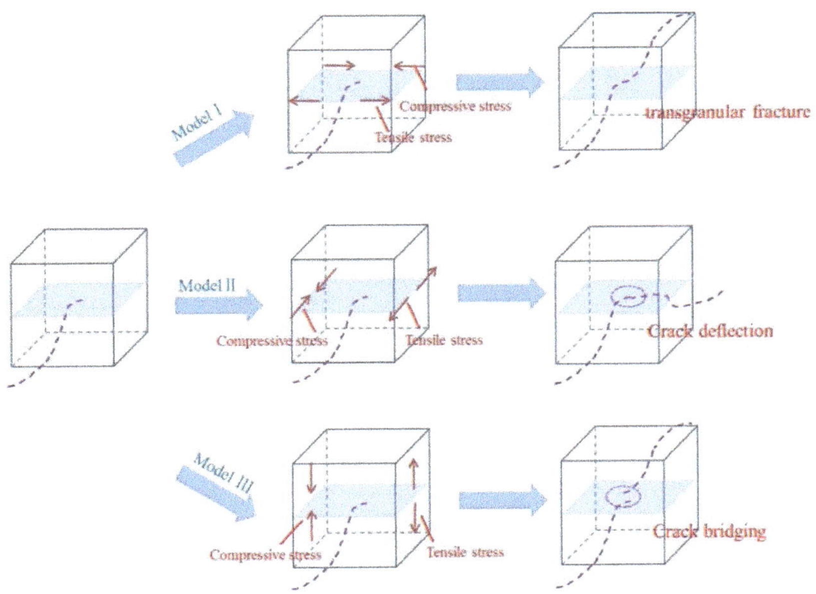

Figure 8. Schematic diagram of crack growth behavior under the synergistic action of tension and compression.

3.3.2. Effect of Interface Binding

Figure 9a shows how the fluorite phase exhibited a transgranular fracture mode, indicating weaker fracture toughness in the fluorite phase. Figure 9b shows that the composites exhibited a crack deflection mode, indicating stronger fracture toughness in the M-$YTaO_4$ phase. When the crack propagated to position P, path 1 appeared along the interface between M-$YTaO_4$ and M-$YTaO_4$ grains and path 2 along the M-$YTaO_4$/fluorite grain boundaries. The existence of path 2 suggests that the interface strength of M-$YTaO_4$/fluorite was weaker than that of M-$YTaO_4$/M-$YTaO_4$.

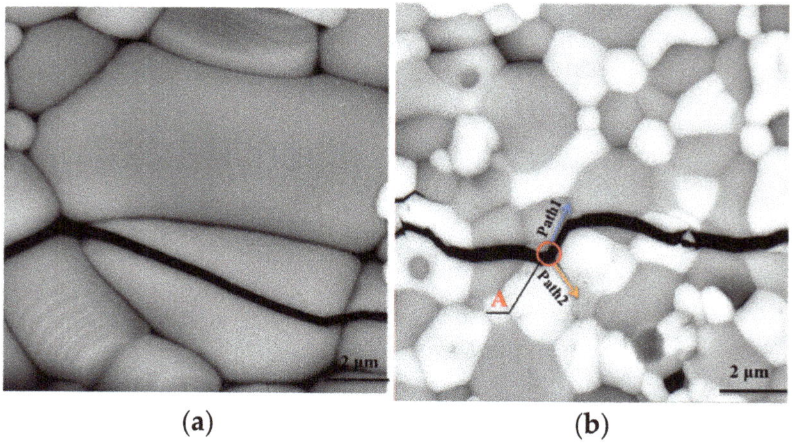

Figure 9. Cracks propagation behavior in YT1 showing a transgranular fracture mode (**a**), and in YT2-YT4 showing an deflection mode triggering mechanism (**b**).

As described above, there were numerous interfaces between M-$YTaO_4$ and Fluorite grains in the composites. Therefore, the first problem was how does the M-$YTaO_4$/fluorite interface strength affect the fracture toughness of the composite? Unfortunately, measuring the interface strength between the grains is almost impossible. In order to prove that the change in doping had an effect on Young's modulus of the composite, a linear analysis was carried out following Voigt's work [37]. Voigt provided a qualitatively analyzed model to investigate how the interface strength affected Young's modulus. The equation for the related calculation is as follows:

$$E_{\text{cal}} = xE_s + (1 - x)E_m \tag{8}$$

where s and x represent the content of the second phase and the second phase, respectively, and m is the matrix phase. Table 4 shows Young's modulus of the composites calculated values and the experimental values.

Table 4. Young's modulus of sample.

Sample	Young's Modulus *E* (GPa)	
	Calculated Values	Measured Values
YT1	203.0	150.8
YT2	196.8	179.8
YT3	190.2	152.9
YT4	183.6	119.9
YT5	177.0	78.3 [*]

[*] Value obtained from the three points loading flexure.

Obviously, Young's modulus is measured by a value lower than the calculated value. However, Voigt believed that strengthening interface bonding meant increasing Young's modulus [26]. Figure 10 shows that the measured value first increased and then decreased, reaching a maximum value at YT2. Combined with Figure 4, the nonlinearity behavior of Young's modulus may have been caused by M-$YTaO_4$ grains, which were wrapped by fluorite phase grains at YT2. The fluorite phase grains and M-$YTaO_4$ grains were closely combined to enhance the interface's strength. However, as the second phase increased, the second phase grain became larger, and the pores increased. This led to the rapid decline of Young's modulus. In addition, both the calculated and measured values of Young's modulus showed a downward trend after adding M-$YTaO_4$. As we know, a low Young's modulus can result in higher strain tolerance in the thermal barrier coating, effectively alleviating the stress caused by thermal shock [38]. Based on the above discussion, we could conclude that M-$YTaO_4$ doping had a significant impact on the interface bonding of composite materials.

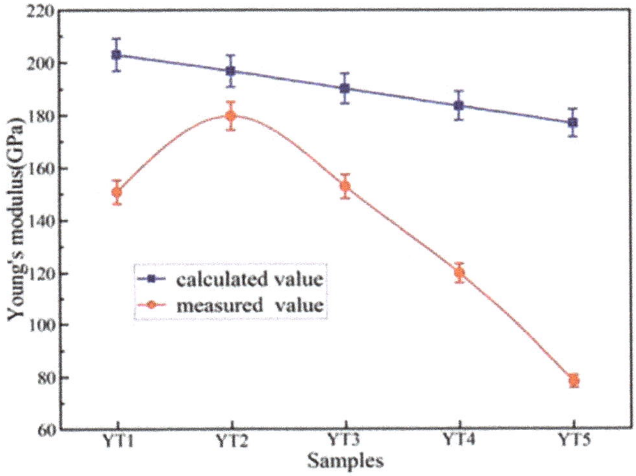

Figure 10. The relationship between the measured value of Young's modulus and the value calculated according to the Voigt model.

3.3.3. Analysis of Crack Propagation Mechanism

In this study, composites YT1 mainly revealed a transgranular fracture mode, as shown in Figure 11a. Cracks were initiated in the matrix and crossed through the fluorite phase grain. The crack extension of the composites YT2-YT4 is shown in Figure 11b–d. When the crack encountered the second phase (M-$YTaO_4$ grains), crack deflection, bridging and bifurcation were the main modes of propagation. The crack needed to consume more energy for propagation because of the higher interface bonding of M-$YTaO_4$ grains, giving rise to an enhanced toughness, which could account for the obviously high toughness of the composite. Thus, cracks deflection, bridging and bifurcation were considered important toughening mechanisms. Hence, introducing the second phase was a primary contributor to the improved toughness of the ceramic materials.

Due to the unfixed angle between the tensile interface and the layer interface, when the crack encountered the M-$YTaO_4$ grain interface, it could penetrate or deflect, adding new free surfaces and releasing more fracture energy. Figure 11 exhibits three fracture modes, crack deflection, intergranular fracture, and crack bridging, indicating that the strength of M-$YTaO_4$ grains was larger than those of the fluorite phase, which led to the toughness of the two-phase region is improved. By analyzing the crack propagation behavior in the composite material, the bonding strength of the two-phase interface could be recognized. Similar phenomena have been observed by other researchers [39].

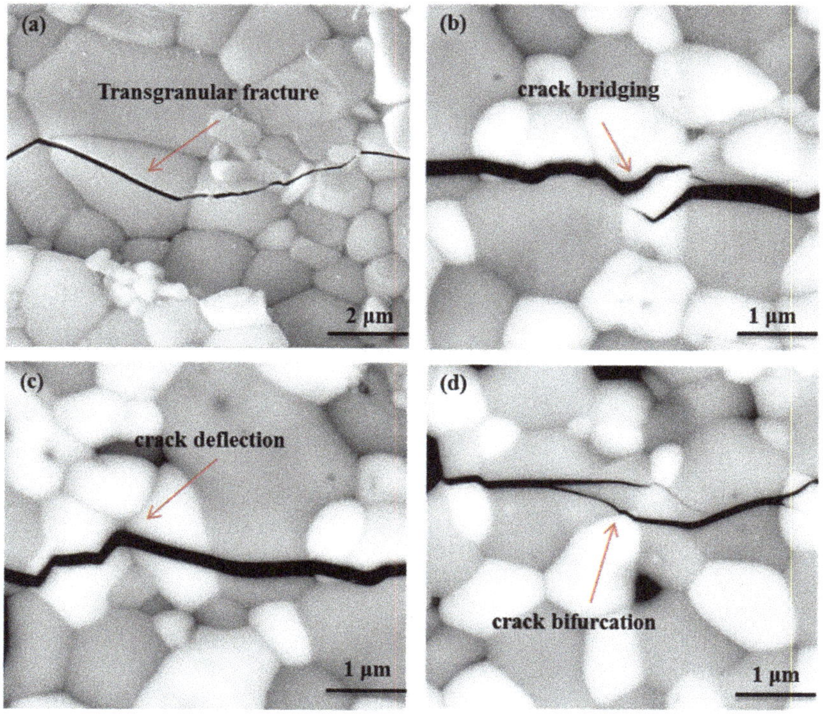

Figure 11. Transgranular fracture in YT1 (**a**), crack bridging (**b**), deflection (**c**), and bifurcation (**d**) in YT2-YT4 due to second phase doping.

3.3.4. Toughening Mechanism of Ferroelastic Domain of $YTaO_4$

M-$YTaO_4$ is a stable ferroelastic structure at room temperature. Due to the T → M phase transition of $YTaO_4$ being ferroelastic, it is of a continuous second-order nature [40]. Previous research states that the T → M phase transition of $YTaO_4$ is ferroelastic, the high symmetry phase (T) breaks into the low symmetry phase (M) at the Curie point, and the corresponding spontaneous strain is:

$$e(S_1) = \begin{pmatrix} e_1 & e_6 & 0 \\ e_6 & e_2 & 0 \\ 0 & 0 & e_3 \end{pmatrix} \quad (9)$$

where:

$e_1 = \frac{c_M \sin\beta}{a_T} - 1$; $e_2 = \frac{a_M}{a_T} - 1$; $e_3 = \frac{b_M}{c_T} - 1$; $e_6 = -\frac{c_M \cos\beta}{2a_T}$.

There are two different orientation states in the monoclinic phase related to the two ferroelastic variants [41,42]. We observed that variants M1 and M2 could switch to each other in the process of crack propagation. The crystallography of the ferroelastic transition and the ferroelastic switching is present in Figure 12, the deformed geometry of M1 was obtained by a tetragonal inclusion undergoing an eigenstrain of $e(S_1)$, and the M2 variant was obtained in the same way, and was associated with the eigenstrain of $e(S_2)$. Alternatively, one could rotate the M1 around the c_T axis 90° clockwise to obtain M2, which is termed ferroelastic switching. To corroborate the above theory, the XRD analysis of the lattice parameters (a b c), volume (V) of $YTaO_4$, and Fluorite phase were obtained, as shown in Table 5.

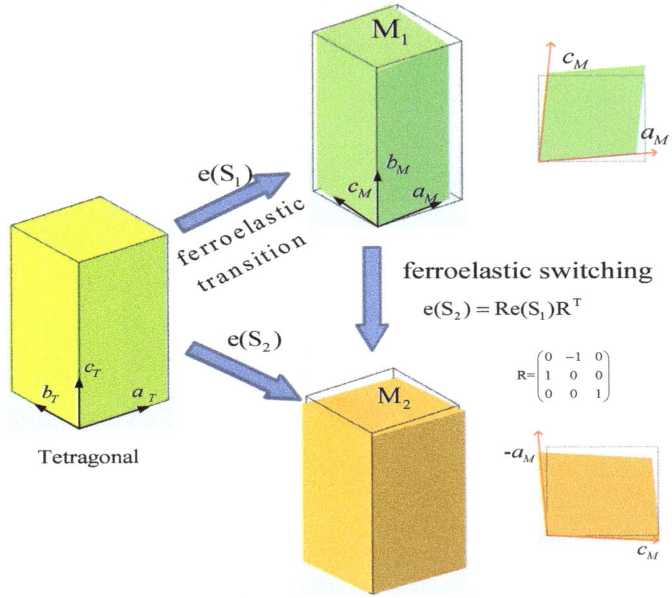

Figure 12. Schematic diagram of ferroelastic transition (T → M) and switching (M1–M2) of $YTaO_4$. The deformed geometries of M1 and M2 were reconstructed using a finite element method. The lattice constants of tetagonal crystal are borrowed from an ab inito calculation [43], and the eigenstrain of Equation (9), associated with a deformed inclusion, (M1) could be derived from the lattice constants of YT5 of Table 5.

Table 5. Calculated lattice parameters (a b c) and volume (*V*) of $YTaO_4$ and Fluorite phase.

Sample	M-$YTaO_4$				Fluorite	
	a (Å)	b (Å)	c (Å)	vol (Å^3)	abc (Å)	vol (Å^3)
YT1	5.326	10.932	5.05	292.65	5.2168	141.98
YT2	5.3189	10.9038	5.0666	292.59	5.2098	141.41
YT3	5.3324	10.9317	5.0423	292.59	5.2106	141.46
YT4	5.3334	10.9337	5.0389	292.5	5.2099	141.42
YT5	5.3252	10.9313	5.0538	292.83	-	

Figure 13a,b shows that the ferroelastic domain existed of parallel striped structures in the pure $YTaO_4$ ceramic (samples YT5) and also indicated that there was always only one polarization direction in one grain when the grain was far from the crack. Furthermore, we found that the ferroelastic structure became more pronounced as the grains became larger. Figure 13c shows that M1 and M2 coexisted in one grain and were nearly perpendicular to each other, which agreed with theoretical studies. Therefore, the M2 variant transformed into the M1 variant and could be seen as a sort of rotating twinning process associated with the rotation described in Figure 12. Ferroelastic domains with the different polarization direction (A and B) were observed near the crack tip in the bridging process, as shown in Figure 13c, which indicated that the factors to trigger ferroelastic switching were not only dependent on the crack issue, but the grain size and grain boundary geometry may have also affected the switching. Ferroelastic toughening was evidenced in Figure 13d when the crack penetrated the grain, and two polarization directions (C and D) were observed.

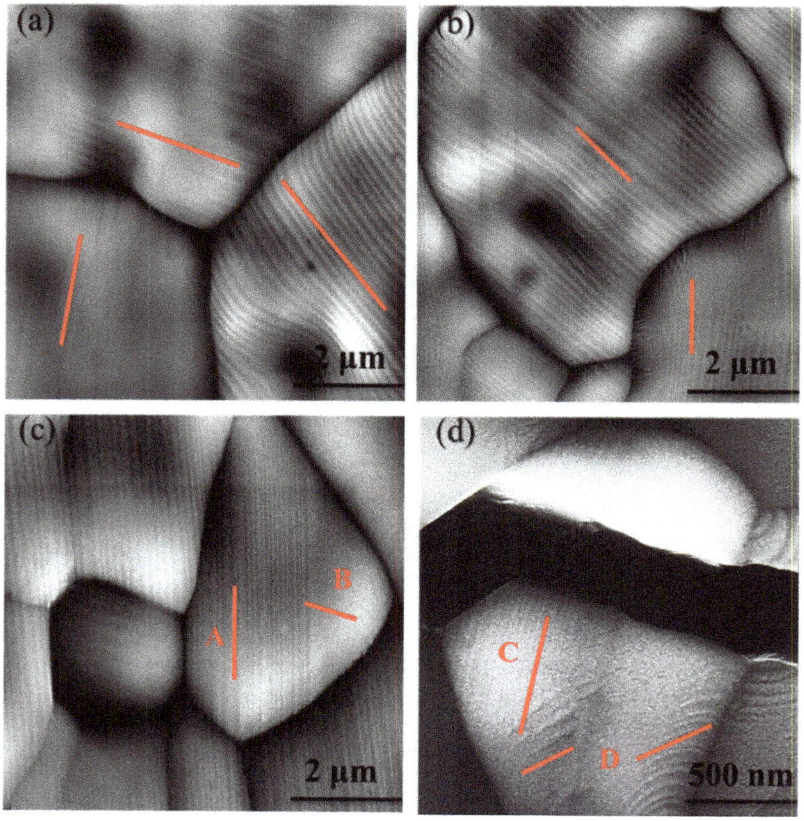

Figure 13. The sample YT5 pure M-YtaO$_4$ ferroelastic domain structure was observed by SEM (**a**,**b**), the twin grain visible in the figure is the M phase resulting from the displacement phase transformation of t-YTaO$_4$ during cooling. Ferroelastic switching in the bridging part of the crack tip in pure YTaO$_4$ (**c**). The sample YT3 is subjected to external stress and the ferroelastic domain structure is transformed (**d**).

As we know, ferroelastic switching can be seen as a sort of mechanical twinning [44]. Twinning stress was influenced by grain size, followed by the Hall–Petch law [45]. An increase in grain size served to decrease the critical twinning stress. In the experiments, we indeed found that, in the larger grain, ferroelastic switching near the crack was more deterministic than in the small grains. Therefore, we believe that ferroelastic toughening occupies an important position compared to others which can affect the toughness of the composite ceramic with an increase in m-YTaO$_4$ doping.

4. Conclusions

In this study, ZYTO composite ceramic materials were prepared by the chemical coprecipitation method. The microstructure of ZYTO was studied, as well as the mechanical properties and toughening mechanisms. The following conclusions can be drawn:

ZYTO composite ceramics show excellent phase stability and no chemical reactions between the two phases. The grain size of the two phases was gradually consistent with increasing the M-YTaO$_4$ doping concentration. The M-YTaO$_4$ phase refined the fluorite phase grain.

Through calculation and analysis, both the density and porosity of the ZYTO composite ceramics increased with increasing doping. The hardness decreased almost linearly with the doping of M-$YTaO_4$, from 10.6 to 4.5 GPa. The fracture toughness first increased and then decreased with the M-$YTaO_4$ doping concentration, and YT3 possessed the highest fracture toughness of 3.1 $MPa \cdot m^{1/2}$.

The residual stress, interface state, crack propagation mechanism and ferroelastic switching had important effects on the toughness of ZYTO composite ceramics. In this case, due to the introduction of M-$YTaO_4$ grains, the toughness of ZYTO composite ceramics was impacted by second-phase toughening and ferroelastic toughening, which helped to further improve the fracture toughness. We believe that the second phase of toughening and ferroelastic toughening played a dominant role in improving the toughness of the material.

Author Contributions: Writing—original draft, X.F.; Writing—review and editing, F.Z., W.Z. and Z.P. All authors have read and agreed to the published version of the manuscript."

Funding: This research is sponsored by National Natural Science Foundation of China (NSFC) under Grant No. 52171015, and Science and Technology Innovation Program of Hunan Province (Grant No. 2022RC1082).

Institutional Review Board Statement: Not applicable.

Informed Consent Statement: Not applicable.

Data Availability Statement: Not applicable.

Conflicts of Interest: The authors declare no conflict of interest.

References

1. Hu, Z.C.; Liu, B.; Wang, L.; Cui, Y.H.; Wang, Y.W.; Ma, Y.D.; Sun, W.W.; Yang, Y. Research progress of failure mechanism of thermal barrier coatings at high temperature via finite element method. *Coatings* **2020**, *10*, 732. [CrossRef]
2. Yuan, B.; Wang, W.; Ye, D.; Zhang, Z.; Fang, H.; Yang, T.; Wang, Y.; Zhong, S. Nondestructive Evaluation of Thermal Barrier Coatings Thickness Using Terahertz Technique Combined with PCA–GA–ELM Algorithm. *Coatings* **2022**, *12*, 390. [CrossRef]
3. Zhou, Y.; Yang, L.; Zhu, W. *Thermal Barrier Coatings: Failure Theory and Evaluation Technology*; Springer Nature: Berlin, Germany, 2022.
4. Parchovianský, M.; Parchovianská, I.; Hanzel, O.; Netriová, Z.; Pakseresht, A. Phase Evaluation, Mechanical Properties and Thermal Behavior of Hot-Pressed LC-YSZ Composites for TBC Applications. *Materials* **2022**, *15*, 2839. [CrossRef]
5. Parchovianská, I.; Parchovianský, M.; Nowicka, A.; Prnová, A.; Švančárek, P.; Pakseresht, A. Effect of Sintering Temperature on Phase Evolution, Microstructure, and Mechanical Properties of $La_2Ce_2O_7$/40 wt.% YSZ Composite Ceramics. *J. Mater. Res. Technol.* **2023**, *24*, 4573–4586. [CrossRef]
6. Liu, X.; Zhang, P.; Han, Y.; Pan, W.; Wan, C. Tailoring thermal and mechanical properties of rare earth niobates by coupling entropy and composite engineering. *J. Eur. Ceram. Soc.* **2023**, *43*, 1141–1146. [CrossRef]
7. Han, Y.; Zong, P.A.; Huang, M.; Yang, Z.; Feng, Y.; Pan, W.; Zhang, P.; Wan, C. In-situ synthesis of gadolinium niobate quasi-binary composites with balanced mechanical and thermal properties for thermal barrier coatings. *J. Adv. Ceram.* **2022**, *11*, 1445–1456. [CrossRef]
8. Macauley, C.A.; Fernandez, A.N.; Van Sluytman, J.S. Phase equilibria in the ZrO_2-$YO_{1.5}$-$TaO_{2.5}$ system at 1250 °C. *J. Eur. Ceram. Soc.* **2018**, *38*, 4523–4532. [CrossRef]
9. Macauley, C.A.; Fernandez, A.N.; Levi, C.G. Phase equilibria in the ZrO_2-$YO_{1.5}$-$TaO_{2.5}$ system at 1500 °C. *J. Eur. Ceram. Soc.* **2017**, *37*, 4888–4901. [CrossRef]
10. Limarga, A.M.; Shian, S.; Leckie, R.M. Thermal conductivity of single-and multi-phase compositions in the ZrO_2-Y_2O_3-Ta_2O_5 system. *J. Eur. Ceram. Soc.* **2014**, *34*, 3085–3094. [CrossRef]
11. Guo, X.; Wang, Z. Effect of niobia on the defect structure of yttria-stabilized zirconia. *J. Eur. Ceram. Soc.* **1998**, *18*, 237–240. [CrossRef]
12. Heinze, S.G. *Phase Equilibria and Toughness of ZrO_2-(Y/Yb)$O_{1.5}$-$TaO_{2.5}$ Thermal Barrier Coatings*; University of California: Santa Barbara, CA, USA, 2018.
13. Han, M.; Tang, X.; Yin, H. Fabrication, microstructure and properties of a YSZ electrolyte for SOFCs. *J. Power Sources* **2007**, *165*, 757–763. [CrossRef]
14. Fünfschilling, S.; Fett, T.; Hoffmann, M.J. Mechanisms of toughening in silicon nitrides: The roles of crack bridging and microstructure. *Acta Mater.* **2011**, *59*, 3978–3989. [CrossRef]
15. Ferraro, C.; Meille, S.; Réthoré, J. Strong and tough metal/ceramic micro-laminates. *Acta Mater.* **2018**, *144*, 202–215. [CrossRef]

16. Bartolomé, J.F.; Gutiérrez-González, C.F.; Pecharroman, C. Synergistic toughening mechanism in 3Y–TZP/Nb composites. *Acta Mater.* **2007**, *55*, 5924–5933. [CrossRef]
17. Kumar, R.S. Crack-growth resistance behavior of mode-I delamination in ceramic matrix composites. *Acta Mater.* **2017**, *131*, 511–522. [CrossRef]
18. Stefan, G.H.; Jason, S.V.S.; Carlos, G.L. Microstructure evolution and physical properties of ZrO_2-(Y+Yb)$O_{1.5}$-$TaO_{2.5}$ thermal barrier coatings. In *Surface and Coatings Technology*; Elsevier: Amsterdam, Netherlands, 2020; Volume 389, p. 125648. ISSN 0257-8972.
19. Raghavan, S.; Wang, H.; Dinwiddie, R.B.; Porter, W.D.; Vaβen, R.; Stöver, D.; Mayo, M.J. Ta_2O_5/Nb_2O_5 and Y_2O_3 co-doped zirconias for thermal barrier coatings. *J. Am. Ceram. Soc.* **2004**, *87*, 431–437. [CrossRef]
20. Kulyk, V.; Duriagina, Z.; Kostryzhev, A.; Vasyliv, B.; Vavrukh, V.; Marenych, O. The effect of yttria content on microstructure, strength, and fracture behavior of yttria-stabilized zirconia. *Materials* **2022**, *15*, 5212. [CrossRef]
21. Kulyk, V.; Duriagina, Z.; Vasyliv, B.; Vavrukh, V.; Kovbasiuk, T.; Lyutyy, P.; Vira, V. The effect of sintering temperature on the phase composition, microstructure, and mechanical properties of yttria-stabilized zirconia. *Materials* **2022**, *15*, 2707. [CrossRef]
22. Evans, A.G.; Charles, E.A. Fracture toughness determinations by indentation. *J. Am. Ceram. Soc.* **1976**, *59*, 371–372. [CrossRef]
23. Asmani, M.; Kermel, C.; Leriche, A.; Ourak, M. Influence of porosity on Young's modulus and Poisson's ratio in alumina ceramics. *J. Eur. Ceram. Soc.* **2001**, *21*, 1081–1086. [CrossRef]
24. Wan, C.L.; Qu, Z.X.; Du, A.B.; Pan, W. Influence of B site substituent Ti on the structure and thermophysical properties of $A_2B_2O_7$-type pyrochlore $Gd_2Zr_2O_7$. *Acta Mater.* **2009**, *57*, 4782–4789. [CrossRef]
25. Standard, A. E112: Standard test methods for determining average grain size. *J. West Conshocken* **1996**, *112*, 4–20.
26. Wang, C.M.; Guo, L.; Ye, F.X. $LaPO_4$ as a toughening agent for rare earth zirconate ceramics. *Mater. Des.* **2016**, *111*, 389–393. [CrossRef]
27. Yang, J.; Wan, C.; Zhao, M.; Shahid, M.; Pan, W. Effective blocking of radiative thermal conductivity in $La_2Zr_2O_7$/$LaPO_4$ composites for high temperature thermal insulation applications. *J. Eur. Ceram. Soc.* **2016**, *36*, 3809–3814. [CrossRef]
28. Ren, X.R.; Guo, S.C.; Zhao, M.; Pan, W. Thermal conductivity and mechanical properties of YSZ/LaPO4 composites. *J. Mater. Sci.* **2014**, *49*, 2243–2251. [CrossRef]
29. Du, A.B.; Pan, W.; Ahmad, K.; Shi, S.L.; Qu, Z.X.; Wan, C.L. Enhanced mechanical properties of machinable $LaPO_4$/Al_2O_3 composites by spark plasma sintering. *Int. J. Appl. Ceram. Technol.* **2009**, *6*, 236–242. [CrossRef]
30. Zhou, H.M.; Guo, Y.J.; Li, J.; Yi, D.Q.; Xiao, L.R. Microstructure and mechanical properties of 8YSZ ceramics by liquid-phase sintering with CuO-TiO_2 addition. *J. Cent. South Univ.* **2012**, *19*, 1196–1201. [CrossRef]
31. Mercer, C.; Williams, J.R.; Clarke, D.R.; Evans, A.G. On a ferroelastic mechanism governing the toughness of metastable tetragonal-prime (t′) yttria-stabilized zirconia. *Proc. R. Soc. A Math. Phys. Eng. Sci.* **2007**, *463*, 1393–1408. [CrossRef]
32. Steinberg, L.; Naraparaju, R.; Heckert, M.; Mikulla, C.; Schulz, U.; Leyens, C. Erosion behavior of EB-PVD 7YSZ coatings under corrosion/erosion regime: Effect of TBC microstructure and the CMAS chemistry. *J. Eur. Ceram. Soc.* **2018**, *38*, 5101–5112. [CrossRef]
33. Taya, M.; Hayashi, S.; Kobayashi, A.S.; Yoon, H.S. Toughening of a particulate reinforced ceramic-matrix composites by thermal residual stress. *J. Am. Ceram. Soc.* **1990**, *73*, 1382–1391. [CrossRef]
34. Zhou, Y.; Gan, M.; Yu, W.; Chong, X.; Feng, J. First-principles study of thermophysical properties of polymorphous $YTaO_4$ ceramics. *J. Am. Ceram. Soc.* **2021**, *104*, 6467–6480. [CrossRef]
35. Chen, L.; Song, P.; Feng, J. Influence of ZrO_2 alloying effect on the thermophysical properties of fluorite-type Eu_3TaO_7 ceramics. *Scr. Mater.* **2018**, *152*, 117–121. [CrossRef]
36. Wang, D.Z.; Hu, Q.W.; Zeng, X.Y. Residual stress and cracking behaviors of $Cr_{13}Ni_5Si_2$ based composite coatings prepared by laser-induction hybrid cladding. *Surf. Coat. Technol.* **2015**, *274*, 51–59. [CrossRef]
37. Chung, D.H. The Voigt-Reuss-Hill (VRH) approximation and the elastic moduli of polycrystalline ZnO, TiO_2 (Rutile), and -Al_2O_3. *J. Appl. Phys.* **1968**, *39*, 2777–2782. [CrossRef]
38. Shen, Z.; Liu, Z.; Mu, R.; He, L.; Liu, G.; Huang, Z. LaGdZrO/YSZ thermal barrier coatings by EB-PVD: Microstructure, thermal properties and failure mechanism. *Chem. Eng. J. Adv.* **2021**, *5*, 100073. [CrossRef]
39. Cao, Y.N.; Li, C.; Ma, Y.; Luo, H.Y.; Yang, Y.H.; Guo, H.B. Mechanical properties and thermal conductivities of 3YSZ-toughened fully stabilized HfO2 ceramics. *Ceram. Int.* **2019**, *45*, 12851–12859. [CrossRef]
40. Shian, S.; Sarin, P.; Gurak, M.; Baram, M.; Kriven, W.M.; Clarke, D.R. The tetragonal–monoclinic, ferroelastic transformation in yttrium tantalate and effect of zirconia alloying. *Acta Mater.* **2014**, *69*, 196–202. [CrossRef]
41. Luo, C.; Li, C.; Cao, K.; Li, J.; Luo, J.; Zhang, Q.; Zhou, Q.; Zhang, F.; Gu, L.; Yang, L.; et al. Ferroelastic domain identification and toughening mechanism for yttrium tantalate–zirconium oxide. *J. Mater. Sci. Technol.* **2022**, *127*, 78–88. [CrossRef]
42. Wang, J.; Zheng, Q.; Shi, X.; Li, D.; Yang, Y.; Li, C.; Feng, J. Microstructural evolution and thermal-physical properties of $YTaO_4$ coating after high-temperature exposure. *Surf. Coat. Technol.* **2023**, *456*, 129222. [CrossRef]
43. Feng, J.; Shian, S.; Xiao, B.; Clarke, D.R. First-principles calculations of the high-temperature phase transformation in yttrium tantalate. *Phys. Rev. B* **2014**, *90*, 094102. [CrossRef]

44. Arlt, G. Twinning in ferroelectric and ferroelastic ceramics: Stress relief. *J. Mater. Sci.* **1990**, *25*, 2655–2666. [CrossRef]
45. Korolev, P.V.; Savchenko, N.L.; Kul'kov, S.N. Texture formation on the friction surface in transformation-toughened ceramics. *Tech. Phys. Lett.* **2004**, *30*, 12–14. [CrossRef]

Disclaimer/Publisher's Note: The statements, opinions and data contained in all publications are solely those of the individual author(s) and contributor(s) and not of MDPI and/or the editor(s). MDPI and/or the editor(s) disclaim responsibility for any injury to people or property resulting from any ideas, methods, instructions or products referred to in the content.

coatings

MDPI

Article

Influence of TiC Addition on Corrosion and Tribocorrosion Resistance of Cr_2Ti-NiAl Electrospark Coatings

Konstantin A. Kuptsov [1], Mariya N. Antonyuk [1], Alexander N. Sheveyko [1], Andrey V. Bondarev [2,*] and Dmitry V. Shtansky [1]

1 National University of Science and Technology "MISIS", Leninsky prospect 4, Moscow 119049, Russia
2 Department of Control Engineering, Faculty of Electrical Engineering, Czech Technical University in Prague, Technicka 2, Prague 6, 16627 Prague, Czech Republic
* Correspondence: bondaan2@fel.cvut.cz

Abstract: Marine and coastal infrastructures usually suffer from synergetic effect of corrosion and wear known as tribocorrosion, which imposes strict requirements on the structural materials used. To overcome this problem, novel composite wear- and corrosion-resistant xTiC-Fe-CrTiNiAl coatings with different TiC content were successfully developed. The coatings were obtained by the original technology of electrospark deposition in a vacuum using xTiC-Cr_2Ti-NiAl (x = 0, 25, 50, 75%) electrodes. The structure and morphology of the coatings were studied in detail by XRD, SEM, and TEM. The effect of TiC content on the tribocorrosion behavior of the coatings was estimated using tribological and electrochemical (under stationary and wear conditions) experiments, as well as impact testing, in artificial seawater. The TiC-free Fe-Cr_2Ti-NiAl coating revealed a defective inhomogeneous structure with transverse and longitudinal cracks. Introduction of TiC allowed us to obtain coatings with a dense structure without visible defects and with uniformly distributed carbide grains. The TiC-containing coatings were characterized by a hardness and elastic modulus of up to 10.3 and 158 GPa, respectively. Formation of a composite structure with a heavily alloyed corrosion-resistant matrix based on α-(Fe,Cr) solid solution and uniformly distributed TiC grains led to a significant increase in resistance to stationary corrosion and tribocorrosion in artificial seawater. The best 75TiC-Fe-CrTiNiAl coating demonstrated the lowest corrosion current density values both under stationary (0.03 $\mu A/cm^2$) and friction conditions (0.8 $\mu A/cm^2$), and was characterized by both a 2–2.5 times lower wear rate (4×10^{-6} mm^3/Nm) compared to AISI 420S steel and 25TiC-Fe-CrTiNiAl and a high fracture toughness.

Keywords: electrospark deposition; composite coatings; wear resistance; corrosion resistance; artificial seawater

1. Introduction

Over the past decades, exploration of marine resources and development of marine industry have grown significantly. However, every year, corrosion of parts and equipment of ships and marine infrastructure leads to their premature failure, which entails significant losses as well as environmental pollution [1].

One of the most widely used materials for marine and coastal infrastructure parts is hardened stainless steel, due to its good corrosion resistance and satisfactory wear resistance [2]. The corrosion resistance of stainless steels is due to alloying with chromium, which leads to the formation of a protective passive chromium oxide film on its surface [3]. However, some loaded parts, such as pumps, valves, gears, shafts, propellers, etc., are subjected to intense friction in an aggressive corrosive environment (such as sea water), which leads to removal of the protective oxide film and, as a result, to a decrease in corrosion and wear resistance [4]. The synergistic effect of friction and corrosion is called tribocorrosion, and results in more severe equipment damage and even higher annual maintenance costs [5].

One of the most effective ways to improve the resistance of marine infrastructure parts to tribocorrosion is deposition of composite coatings consisting of a corrosion-resistant metal matrix and a hardening phase based on intermetallic compounds or transition metal carbides, which may increase the durability, reliability and service life of parts during their operation in harsh sea conditions [6,7]. As a corrosion-resistant matrix, it is preferable to use alloys with a high chromium content that are close in composition to stainless steel (Cr (12–18 wt %)) [8]. Rocha A.M.F. et al. demonstrated that the addition of Cr to the WC-FeCoNi coating significantly improved corrosion resistance [6]. Alizadeh, M et al. [7] investigated the influence of Cr on the corrosion behavior of the NiAl intermetallic compound. The increase in Cr content led to the formation of a denser Cr-based protective oxide layer on the surface which prevented formation of pitting corrosion.

In recent years, intermetallic alloys and coatings such as Ni-Ti, Ni-Al and Ti-Al have been widely used in the marine, oil, and aviation industries due to their high hardness, corrosion resistance and oxidation resistance [9,10]; moreover, these coatings have shown good tribological properties [11]. A significant disadvantage of such coatings is their high brittleness and low crack resistance [12,13], which limits their use in aggressive corrosive environments upon friction or impact loads.

One promising way to overcome this problem is to produce composite materials reinforced by elements or components with high hardness [14]. Coatings based on carbides [15,16], nitrides [17,18] and carbonitrides [19] of transition metals, such as TiC, TaC, WC, TaN, and Ti(C, N), are widely used to improve surface properties under wear and corrosion conditions, due to their high hardness, low friction coefficient and high chemical stability. Many authors have demonstrated that the introduction of such hard particles in corrosion-resistant metal matrices leads to the enhancement of hardness and strength [20], wear and erosion resistance [21], and corrosion resistance [22]. However, choosing the right hardening particles for each matrix is crucial for the corrosion behavior of the composites in terms of matching the corrosion potentials of the reinforcing particles and the metal matrix [23]. For example, the introduction of NbC (15 vol.%) particles into AISI 440 steel was shown to increase its sliding and erosion resistance, while slightly deteriorating its corrosion resistance [21]. Composite high-entropy coatings of CoCrFeNi reinforced with WC particles (up to 40 wt.%) exhibited superior wear resistance at an optimal WC content; meanwhile, a slight decrease in corrosion resistance was observed [24], and TiC-doped CoCrNiFeAl demonstrated superior corrosion resistance (1×10^{-6} A/cm^2) compared to Ti_6Al_4V substrate (1.4×10^{-4} A/cm^2) in 3.5% NaCl [25]. On the contrary, introduction of WC into a Co-Cr matrix led to a significant decrease in corrosion current density, from 3×10^{-3} to 4×10^{-4} A/cm^2, compared to non-modified Co-Cr matrix, due to the pronounced passivation effect of WC particles. However, corrosion potential decreased from 350 to 70 Mv [22].

Titanium carbide (TiC) is widely used to improve the corrosion and wear resistance of composite materials, as a hardening phase [26,27]. In our work, TiC was chosen because on the one hand, its corrosion potential is high enough to ensure effective passivation of Cr-containing metal matrix, but on the other hand, its corrosion potential is not enough to lead to pitting breakdown of the steel substrate.

Protective coatings for parts of coastal and marine infrastructure are deposited by a number of techniques such as electrospark deposition (ESD) [23,28], PCD and CVD [29,30], laser cladding [31,32], HVOF [14,33], and electroplating [34].

Original technology allowing automated electrospark deposition in a vacuum is a promising method to produce wear- and corrosion-resistant coatings on various steels and titanium alloys [35]. ESD in a vacuum allows the deposition of thick coatings (up to 200 μm) with high adhesive strength, due to the microwelding effect between the substrate and electrode material. In addition, this method is characterized by low cost, relative simplicity, and easy automation of the process, and provides the possibility of local treatment of large parts. During ESD in a vacuum, two parallel processes are implemented: classical mass transfer of yjr material from the electrode to the substrate, and pulsed cathode-arc evaporation of the electrode, initiated by spark breakdowns. This feature increases the efficiency of the process, provides additional opportunities for doping the coating due to the use of composite cathodes, and also contributes to better surface wettability which ensures lower coating roughness. [36].The aim of this work was to obtain novel composite coatings based on a corrosion-resistant Fe-Cr_2Ti-NiAl matrix reinforced with varied content of TiC using the original technology of electrospark deposition in a vacuum, and to study the effect of TiC content on corrosion and tribocorrosion behavior in artificial seawater.

2. Experimental Details

2.1. Electrode Production

Electrodes xTiC-CrTiNiAl (x = 0, 25, 50, 75%) were manufactured from high purity Ti (PTM-1, 5–15 μm), Cr (PKh-1S, <63 μm), and NiAl (PM-NYu50, <40 μm, purity > 99%) powders that were mixed in the following atomic ratios (for x = 0): 26.7% Ti—53.4% Cr—20% NiAl. For TiC-containing electrodes, 25, 50 and 75 at.% of TiC powder was added to the initial mixture. Then, the mixtures were treated in a ball rotary mill with a drum volume of 1 L, at a mixture to balls mass ratio of 1:6 for 2 h.

Electrode bars (7 × 7 × 50 mm) were prepared by cold-pressing using a Lab Econ 600 hydraulic press (Fontijne Grotnes B.V., the Netherlands) at a pressure of 3.5 t/cm^2, and then were sintered in a high-temperature vacuum furnace VE-3-16 at a pressure of 1×10^{-3} Pa and a temperature of 1310 °C for 60 min. After sintering, the electrodes were ground to obtain rods with a diameter of 6 mm.

2.2. Coatings Deposition

The coatings were produced by a pulsed vacuum electrospark deposition (ESD) technology described elsewhere [37].

In this technology, a desktop 3D-processing CNC machine (Technologiya LLC, Russia), in which the cutter drive is replaced by a rotating electrode unit, is installed in the vacuum chamber of the UVN2 m unit. The voltage to the electrode is supplied through an insulated brush assembly. The deposition of coatings was carried out with the following constant parameters: electrode rotation speed 800 rpm, electrode movement speed 500 mm/min, surface scanning step 0.5 mm, number of passes 10 times, electric pulse frequency 100 Hz, pulse voltage 50 V and pulse duration 50 μs. The electrode with respect to the substrate was the cathode. Before ESD, the vacuum chamber was evacuated to a pressure of 10^{-3} Pa. Processing was carried out in an argon atmosphere at a pressure of 0.5 Pa.

Discs made of corrosion-resistant steel AISI 420S (ø30 × 4 mm) with roughness R_a of 6.5 μm were used as substrates.

2.3. Structure and Elemental Composition

The structure, elemental and phase composition of the electrodes and coatings were studied by scanning electron microscopy (SEM) using an S-3400N microscope (Hitachi, Tokyo, Japan) equipped with a NORAN energy-dispersive detector, and X-ray phase analysis (XRD) using a D8 Advance diffractometer (Bruker, Billerica, MA, USA) operating in a Bragg–Brentano geometry with CuKα radiation. Cross-sectional samples for SEM were mechanically cut using Secotom-50 (Struers, Denmark) and mechanically polished using Rotopol-21 (Struers, Denmark) to a roughness R_a of 20 nm. No chemical or ion etching was carried out.

Samples for TEM were produced using a FEI Helios 660 dual-beam system (Helios, NanoLab 660, FEI, USA) with a Ga^+ ion source after depositing a protective W mask. The acceleration voltage for the FIB fabrication was at 30 kV for rough operations, followed by 5 and 2 kV for final thinning. The FIB lamellae were examined using a FEI Titan Themis S/TEM microscope operating (Thermo Fisher, USA) at 300 kV. Using an energy-dispersive

X-ray (EDS) Super-X Detection System with 4 SDD windowless detectors, elemental mapping of the samples was performed.

2.4. Tribocorrosion, Electrochemical and Mechanical Properties Studies

Tribocorrosion investigations of coatings and a steel substrate were carried out in artificial seawater (ASTM D 1141-98) using a tribometer (CSM Instruments, Neuchatel, Switzerland) equipped with a rotating three-electrode cell, which allowed electrochemical measurements to be carried out both under stationary conditions and during friction. This method allows effective estimation of tribocorrosive behavior of both coatings and bulk materials [38,39]. Electrochemical studies were performed using an auxiliary Pt electrode and an Ag/AgCl reference electrode. Potentiodynamic measurements both under stationary conditions and under friction conditions were carried out using a VoltaLab PST050 potentiostat (Radiometer Analytical, France), with a coating surface polarization from −200 to +700 mV relative to the open-circuit corrosion potential at a scanning rate of 1 mV/s. All potentials were recalculated relative to the standard hydrogen electrode. Corrosion current density values were calculated using Tafel equations.

Tribocorrosion studies were carried out using "ball-on-disc" scheme at a load of 5 N, a radial sliding velocity of 10 cm/s, and a run distance of 500 m. An Al_2O_3 ball with a diameter of 6 mm was used as a counterpart. Wear track profiles were studied on a Wyko-NT1100 optical profiler (Veeco, Plainview, NY, USA).

To obtain statistics, all corrosion and tribocorrosion studies were carried out using two different samples of each coating. The measurement errors did not exceed 10%.

The fatigue strength of coatings was assessed using an Impact tester (CemeCon, Germany). The tests were carried out for 10^5 cycles at a constant frequency of 50 Hz and a load of 500 N using a 5 mm in diameter WC-6Co ball. Each sample was tested in air and in artificial seawater.

Mechanical properties (hardness (H) and Young's modulus (E)) were evaluated by nanoindentation and microindentation.

Nanoindentation measurements were carried out on polished cross-sections of coatings using a TI 950 (Hysitron, Billerica, MA, USA). Before measurements, the instrument was calibrated against fused quartz. The measurements were performed in a load control mode, and maximum indentation load was set at 10 mN. The measurements' error did not exceed 10%. The loading and unloading rates were 1 mN/s, while the holding time was 2 s during nanoindentation.

Microindentation was performed on the surface of coatings using an automatic microhardness tester DuraScan 70 (EMCO-TEST Prüfmaschinen GmbH, Austria) by calculating the average value of five measurements. The maximum indentation load was set at 1 N.

3. Results and Discussion

3.1. Structure and Elemental Composition of Electrodes

Figure 1 shows SEM images of the cross-sections of Cr_2Ti-NiAl and 50%TiC-Cr_2Ti-NiAl electrodes, corresponding elemental distribution maps, as well as their composition. For both electrodes, an uneven distribution of metal elements is observed; however, TiC grains of 2–8 μm in size are distributed quite uniformly. The chemical composition of the electrodes, determined by EDS, was close to that of the initial powder mixtures.

3.2. Structure and Elemental Composition of Coatings

3.2.1. SEM

Figure 2 shows SEM images of the surface and cross-sections of the coatings with different content of titanium carbide phase (Figure 2a,b). The coatings were designated as 0TiC, 25TiC, 50TiC and 75TiC according to the content of the carbide phase in the electrodes.

The 0TiC coating deposited with a CrTi-NiAl electrode reveals many defects in the form of a cracks network (Figure 2a). Moreover, these cracks run through the entire coating thickness to the substrate (Figure 2b). The TiC-doped coatings were characterized by a defect-free and uniform morphology without noticeable pores and cracks. The thickness of all coatings did not differ much, and was in the range of 43–45 μm.

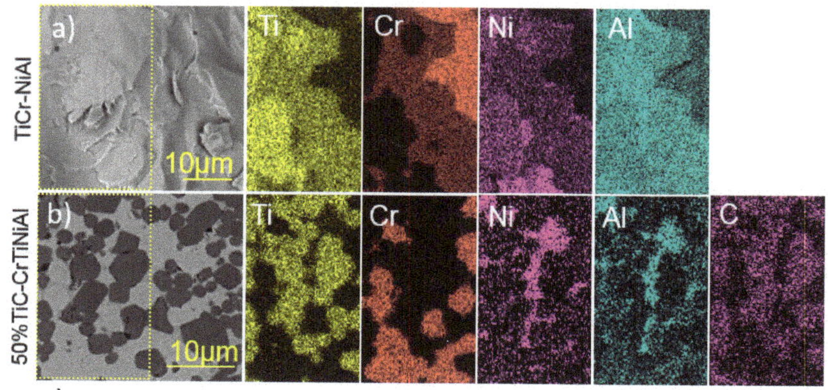

c)

Electrode	Ti	C	Cr	Ni	Al
TiCr-NiAl	26	-	42	15	17
50%TiC-CrTiNiAl	40	27	28	3	2

Figure 1. SEM images of TiCr-NiAl (**a**) and 50%TiC-CrTiNiAl (**b**) electrodes' cross-sections with corresponding EDS element distribution maps and compositions of marked areas (**c**).

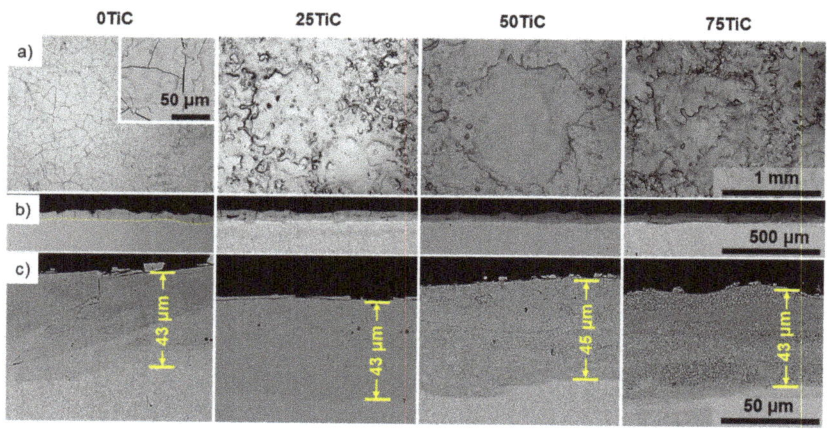

Figure 2. SEM images of surface (**a**) and cross-sections (**b**,**c**) of the coatings.

The 0TiC coating also reveals the lowest Fe content reached only 20 at.%, while carbide-modified coatings demonstrated much higher Fe content, increasing from 45 (75TiC) to 58 at.% (25TiC) (Table 1). This indicates that the introduction of carbides into the electrodes drastically changes mass transfer during the ESD process. When processing with a metallic (Cr_2Ti-NiAl) electrode, predominant electrode material transfer is observed, while in case of treatment with TiC-containing electrodes, electrode and substrate materials are mixed together in approximately equal proportions.

Table 1. Elemental composition of coatings determined by EDS (at.%).

-	C	Ti	Cr	Fe	Ni	Al
0TiC	-	18	39	20	11	12
25TiC	5	11	23	58	1	2
50TiC	9	20	21	47	1	2
75TiC	14	22	17	45	1	1

Higher magnification cross-sectional SEM images of carbide-containing coatings (Figure 3) show that the 25TiC coating is characterized by a small amount of spherical TiC particles that are 100 nm in size. An increase in the amount of carbide phase leads to a significant increase in the number and size of TiC grains (up to 200 nm) in the 50TiC coating. In some areas of the 75TiC coating, formation of dendrites up to 700 nm in size is observed, while the size of most grains do not exceed a size of 200 nm.

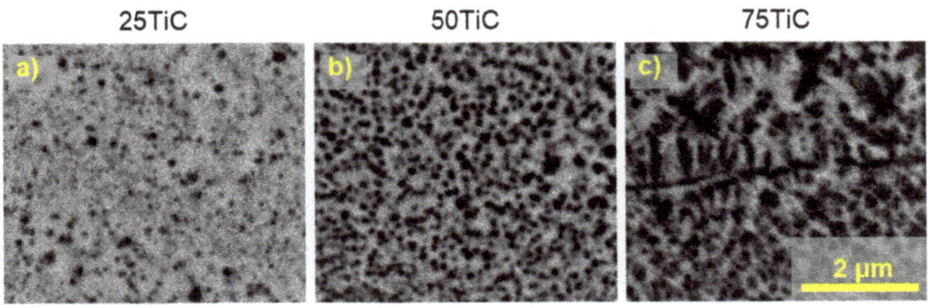

Figure 3. SEM images of cross-sections of 25TiC (**a**), 50TiC (**b**), and 75TiC (**c**) coatings at a higher magnification.

3.2.2. XRD

XRD patterns of the 0TiC, 25TiC, 50TiC, and 75TiC coatings are shown at Figure 4.

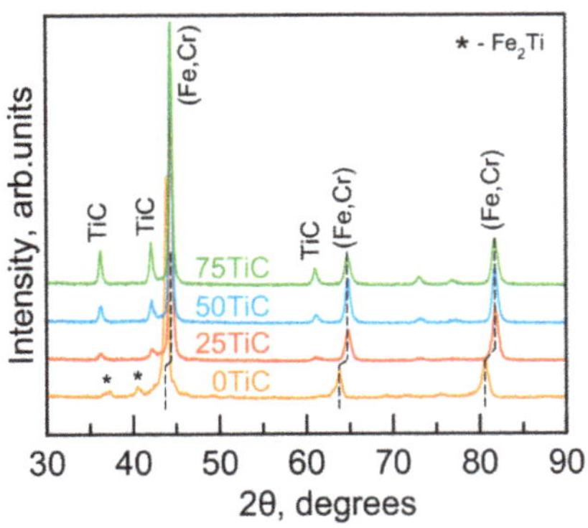

Figure 4. XRD patterns of the coatings.

The XRD pattern of the 0TiC coating has two sets of peaks. The three most intense peaks correspond to (110), (200) and (211) reflexes of a *bcc* Fe-based solid solution with a high content of alloying elements, which is evidenced by a noticeable shift of the peaks towards smaller angles (higher d-spacings). The second set of peaks corresponds to reflections from the (110) and (103) planes of the Fe_2Ti Laves phase (ICDD No. 65-0602).

XRD patterns of the carbide-containing coatings also reveal peaks from *bcc* Fe-based solid solution; however, their positions are typical of those of α-Fe and (Fe,Cr) alloys (ICDD No. 89-4186) [40]. Peaks located at around 36.2°, 42°, and 61° 2θ degrees are attributed to reflexes from (111), (200) and (220) planes of *fcc* TiC (ICDD No. 32-1383). Their intensities gradually increase with an increase in the carbide phase content of the coatings.

3.2.3. TEM

To study the coating's structure in more detail, the 75TiC coating was additionally investigated by TEM.

General views of the coating with corresponding EDS elemental distribution maps are shown in Figure 5. Carbide grains 50–200 nm in size and all elements composing the metal matrix are uniformly distributed over the coating, which agrees well with the SEM studies. It should be noted that the TiC grain size is significantly lower than that of the initial powder. This is due to the fact that carbide grains are not transferred from the electrode during ESD treatment, but precipitate from the melt formed during spark breakdowns. The content of the carbide phase in the coating was determined from the EDS elemental composition and did not exceed 30–40%.

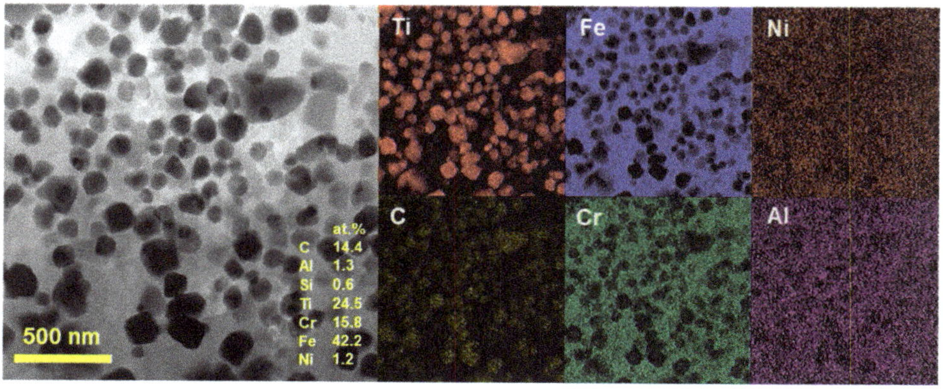

Figure 5. HAADF TEM image of 75C coating with corresponding EDS elemental maps and composition obtained from the whole area.

A number of regions including coarse TiC grains up to 500 nm in size were also observed (Figure 6). These regions were characterized by several features. Due to the high carbide phase content, the size of the matrix grains was limited to 0.5–1 μm (Figure 6a, DF). When analyzing the EDS distribution maps, a higher content of titanium was found at the grain boundaries of the matrix (Figure 6b). It was also revealed that the interfaces are characterized by lower chromium and higher nickel contents, while the iron content at the grain boundaries and in the matrix is similar. Each structural element (marked at Figure 6b as 1, 2 and 3) was additionally investigated using HR TEM and SAED methods. Dark grains (area 1) are attributed to *fcc* TiC carbide. Area 2 corresponds to (Fe, Cr) solid solution with *bcc* structure and d-spacing of 0.204 nm (110). Interestingly, Ti-rich grain boundaries were attributed to Fe_2Ti Laves phase (Figure 6e). It seems that at a small content of titanium ($Ti/(Ti + Fe) < 0.3$) and upon the cooling of the melt, an eutectic transformation occurs with the formation of α-Fe + Fe_2Ti mixture. Apparently, most of the Fe_2Ti was formed at a distance of at least 100–200 nm away from the carbide grains, because in their vicinity, all titanium was absorbed during TiC precipitation. Since the Fe_2Ti phase was observed in XRD pattern only for 0TiC coating, it may be assumed that the amount of that phase in the TiC-containing coatings is negligible due to a lower Ti content and a changed solidification mechanism.

3.3. Mechanical Properties

Figure 7 shows the hardness, H, and elastic modulus, E, values over the coatings' thickness, as well as their corresponding microhardness values measured on the surface of the coatings. Their nanohardness was measured on the cross-sections of the coatings at an angle to the interface from the substrate to the coating. The dashed line in Figure 7a represents the interface between the substrate and coating, while the X-axis corresponds to the distance from that interface. The hardness of the steel substrate was 5 GPa; however, for the 0TiC coating, this value was higher (approximately 7 GPa) due to the extended transition zone. As the distance from the substrate increased, the hardness and elastic modulus increased, reaching their maximum values closer to the surface. For coating 25TiC, the hardness and elastic modulus values exhibited a plateau almost throughout the entire sample, while coatings with higher carbide content demonstrated a gradual increase in H and E values; this can be attributed to the graded TiC content in the coating increasing towards the surface. Interestingly, a maximum hardness and elastic modulus of 11.3 and 178 GPa, respectively, were observed for the 0TiC coating. Coatings with titanium carbide addition were characterized by slightly lower mechanical properties. As the content of carbide phase increased in a row of coatings 25TiC → 50TiC → 75TiC, their hardness and elastic modulus gradually increased from 8.1 to 10.3 GPa, and from 140 to 158 GPa, respectively.

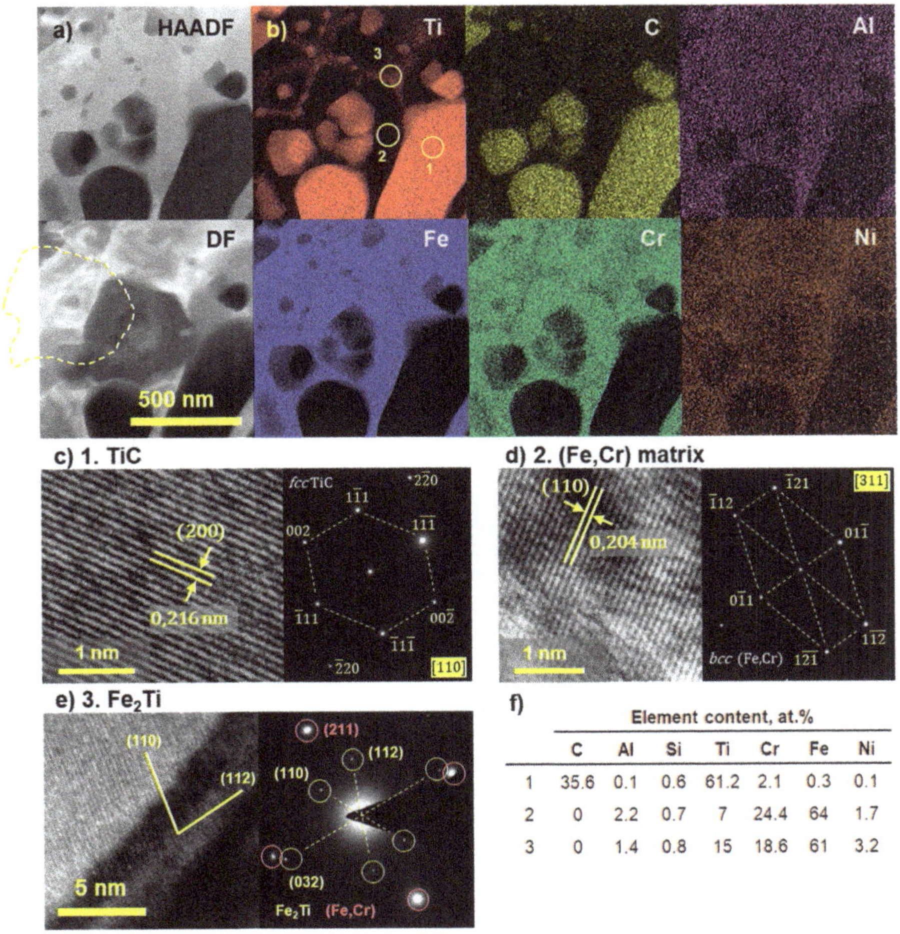

	Element content, at.%						
	C	Al	Si	Ti	Cr	Fe	Ni
1	35.6	0.1	0.6	61.2	2.1	0.3	0.1
2	0	2.2	0.7	7	24.4	64	1.7
3	0	1.4	0.8	15	18.6	61	3.2

Figure 6. HAADF and DF TEM images of 75TiC coating (**a**) with EDS elemental maps (**b**) and corresponding HR TEM images, SAED patterns, and elemental composition (**f**) of marked areas 1 (**c**), 2 (**d**), and 3 (**e**).

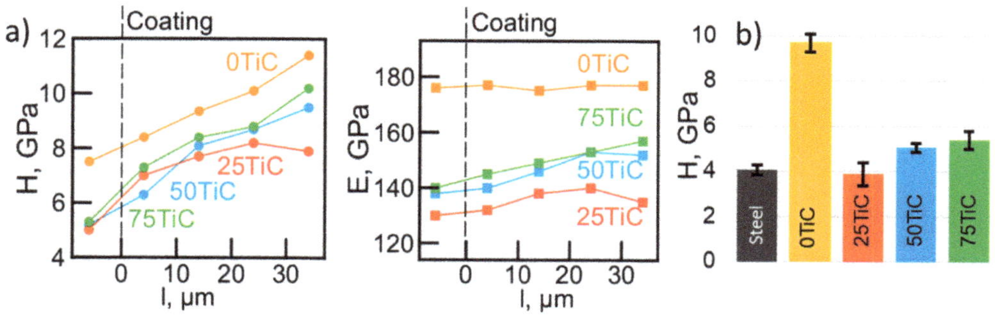

Figure 7. Results of nanoindentation (hardness, H, and elastic modulus, E) (**a**) and microhardness measurements (**b**).

The results of microhardness studies are shown in Figure 7b. The trend in microhardness corresponded to that of nanohardness (Figure 7a). A maximum hardness of 9.8 ± 0.6 GPa was observed for the 0TiC coating, while the hardness of the 25TiC, 50TiC, and 75TiC coatings gradually increased by 4 ± 0.6, 5.2 ± 0.2, and 5.5 ± 0.5 GPa, respectively. The hardness of the 420S steel substrate was 4.1 ± 0.1 GPa. The measured microhardness values are typical of those observed in TiC-reinforced composites (up to 3.6 GPa) [41]; however, coatings 50TiC and 75TiC demonstrated even higher values.

The maximum hardness observed for the 0TiC coatings without a carbide phase is probably due to a hardening caused by Fe_2Ti intermetallic inclusions [42] and a solid–solution strengthening of metal matrix [43], which has a greater effect on the hardness of coatings than inclusion of the carbide phase.

3.4. Tribocorrosion and Electrochemical Behavior of Coatings in ASW

For a detailed study of the effect of carbide phase content in coatings on their tribocorrosion behavior in ASW, several groups of experiments were carried out: (1) tribological experiments with the measurement of coefficient of friction (CoF) and open-circuit corrosion potential (OCP), (2) electrochemical experiments under stationary conditions (without friction) and (3) tribological experiments combined with electrochemical measurements (the recording of potentiodynamic curves under friction conditions).

Figure 8 shows the CoF and OCP dependences of the coatings on the running distance during tribocorrosion experiments in artificial seawater.

Before the experiment, the coatings were kept in ASW for 30 min to stabilize the OCP. After exposure, the coatings' potentials stabilized in the range of +80 (0TiC) to +190 mV (75TiC). There is a clear relationship between an increase in carbide phase content in the coatings and an increase in OCP values. This is due to a more positive potential of the carbide phase [44] and, accordingly, a more effective passivation of the surrounding metal matrix.

After exposure, the tribological part of the experiment began with simultaneous CoF and OCP values' recording. Immediately after the start of friction, a sharp drop in the potential to −100 (75TiC), −180 (50TiC) and −280 mV (25TiC) was observed, after which there was a gradual potential decrease to values of −245, −290 and −335 mV, respectively. For the TiC-free coating, a sharp drop in potential to −500 mV was observed at the initial stage, after which the potential stabilized at −300 mV until the end of the experiment. It should be noted that potential of the steel substrate during friction also reached −300 mV. Increasing the TiC content in the coatings also affects CoF and wear-rate values. Coating 0TiC demonstrated the highest CoF and wear-rate values of 0.45 and 1×10^{-5} mm^3/Nm, respectively, while 75TiC coating revealed the best tribological performance, reaching CoF and wear-rate values of 0.35 and 4×10^{-6} mm^3/Nm. Apparently, there is a clear relationship between TiC phase content and tribocorrosion behavior. Increasing the carbide phase content leads to an enhancement of mechanical properties, which has a positive effect on the wear resistance of the coatings. Uniformly distributed TiC grains protect the metal matrix and passive layer from wear, which explains the higher potential values of the 50TiC and 75TiC coatings during friction. Additionally, TiC grains possess higher potential compared to those of the matrix, and promote matrix passivation [38]. A higher TiC grains content leads to more effective repassivation of the coatings after passive film removal during friction.

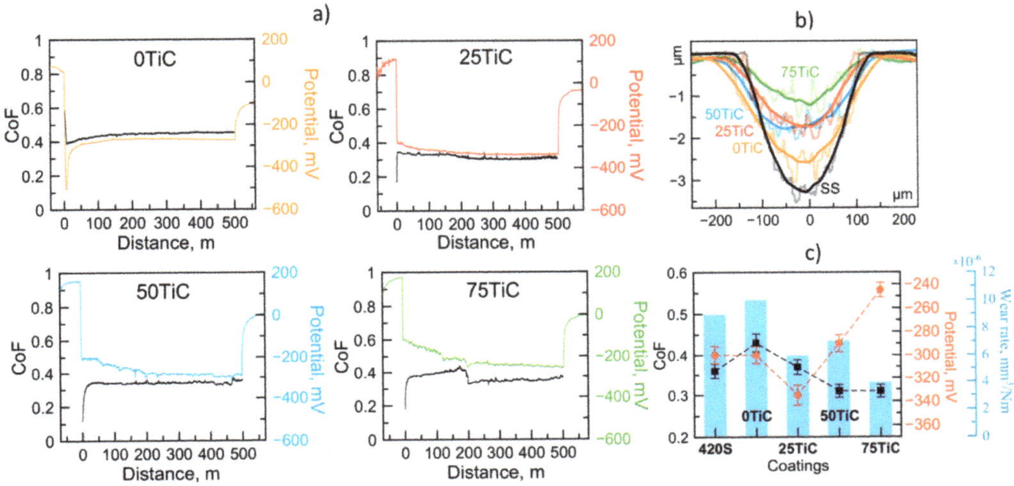

Figure 8. CoF and OCP values vs. distance of tribocorrosion tests in ASW (**a**), corresponding wear tracks profiles (**b**), and a plot summarizing CoF, average wear rates and potential values during tribological tests in ASW (**c**).

Figure 9 shows the results of electrochemical experiments on the coatings and steel substrate, performed both in steady-state and under wear conditions in ASW.

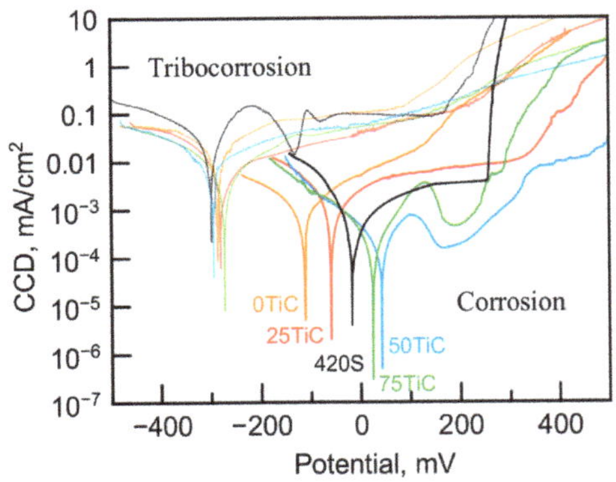

Figure 9. Potentiodynamic curves of coatings and steel substrate recorded during experiments in steady-state (corrosion) and under wear conditions (tribocorrosion).

The polarization curves, corrosion potentials and corrosion current density (CCD) values of the coatings obtained in the absence of friction (steady state) differ greatly (Table 2). First, zero current potentials are more negative than OCP. This is due to the relatively low rate of passive film formation, as there is insufficient time to form a dense layer as the potential shifts from the imposed initial cathode value. The maximum difference between these values is observed for the coating without titanium carbide (as well as for steel), and as the content of carbide increases, this difference decreases. The most negative zero current potential is observed for the 0TiC coating, while the most positive values correspond to coatings 50TiC and 75TiC. The anodic branches of the polarization curves of the 50TiC and 75TiC coatings are characterized by a current density peak in the potential range of 100–200 mV, which is associated with the competing processes of dissolution and isolation on the surface of TiC particles by oxide.

Table 2. The corrosion potential (E_{corr}) and corrosion current density (CCD) of coatings and steel substrate measured during potentiodynamic experiments in steady-state (corrosion) and under wear conditions (tribocorrosion).

-	Corrosion		Tribocorrosion	
Material	**E_{corr}, mV**	**CCD, μA/cm^2**	**E_{corr}, mV**	**CCD, μA/cm^2**
Substrate	-25 ± 3	0.41 ± 0.06	-300 ± 26	22 ± 3
0TiC	-110 ± 6	0.22 ± 0.05	-288 ± 17	9 ± 1
25TiC	-62 ± 7	0.05 ± 0.02	-300 ± 17	6.2 ± 1
50TiC	42 ± 5	0.05 ± 0.02	-282 ± 16	4 ± 0.3
75TiC	24 ± 4	0.03 ± 0.01	-276 ± 9	0.8 ± 0.1

All the coatings do not exhibit a pronounced effect of passive film breakdown (pitting) which was observed on a steel substrate at a potential of +250 mV. Comparison of CCD shows that the metal matrix of the coatings has a slightly higher corrosion resistance compared to that of steel substrate (0.2 vs. 0.4 μA/cm^2, respectively), while the introduction of TiC into the coating composition significantly decreases CCD values to 0.05 (50TiC) and 0.03 μA/cm^2 (75TiC), demonstrating noticeable increase in the corrosion resistance. The measured CCD values are significantly lower than those up to 10 μA/cm^2, which are typically observed for TiC-based composites [41]. The anodic polarization, however, shows insufficient corrosion resistance for 0TiC and 25TiC coatings compared to steel, at least up to the breakdown potential of steel. Coatings with a high TiC content (50TiC and 75TiC) demonstrate low CCD values over the entire potential range.

Figure 10 demonstrates SEM images of wear tracks after tribocorrosion experiments. All wear tracks exhibit similar morphology but differ slightly in width, which is in consistent with the measured wear track profiles (Figure 8b). Dark pits observed both inside and outside the wear tracks are from the initial roughness of the coatings. The pits inside the wear tracks are filled with wear and corrosion products consisting mostly of Fe and Cr oxides (Figure 10e–g).

The results of combined tribological and electrochemical experiments are shown in Figure 11. Potentiodynamic measurements were carried out immediately after stabilization of the OCP and CoF at a distance of 200 m while continuing the friction. During friction, CCD values increase drastically for all the coatings by an average of 20 times (Table 2). However, the current density still tends to a decrease as the content of carbides in coatings increases. The CCD values of substrate and 0TiC coatings increased up to 22 and 9 μA/cm^2, respectively, while that of the 75TiC coating raised from 0.03 to only 0.8 μA/cm^2, evidencing the superior tribocorrosion behavior of the 75TiC coating. Zero current potential values are shifted to the negative side for all coatings, and fall within the range of −276 to −300 mV (Figure 9, Table 2). As in case of stationary electrochemical experiments, no pitting breakdown was observed during friction for all the coatings except for the steel substrate.

It worth noting an interesting feature observed for all the coatings. Under cathodic polarization, CoF is gradually increases and then reaches its maximum values (0.4–0.45) at potentials close to the transition from cathodic to anodic polarization. When the potential is shifted to more positive values, the CoF of all coatings decreases down to 0.25–0.3, which can be explained by an intense oxidation of the surface and a higher content of oxides in the wear debris. The observed effect of reducing CoF during anodic polarization is a very appealing way to increase the wear resistance of coatings capable of passivation.

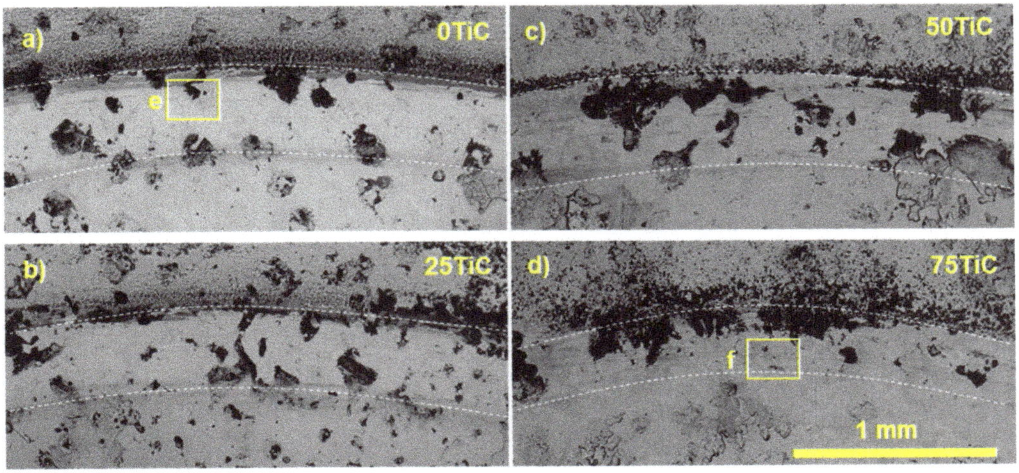

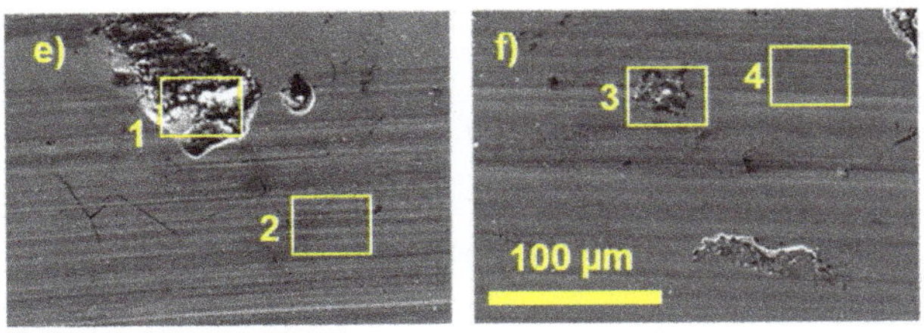

g)	O	Ti	C	Fe	Cr	Ni	Al
1	54	8	2	9	18	4	5
2	5	18	1	22	41	6	7
3	48	13	10	18	9	1	1
4	9	23	17	33	16	1	1

Figure 10. SEM images of the wear tracks on the surface of 0TiC (**a**,**e**), 25TiC (**b**), 50TiC (**c**), and 75TiC (**d**,**f**) coatings after tribocorrosion tests in ASW, as well as corresponding EDS data of marked areas (**g**).

Figure 12 shows SEM images of wear craters on coatings after impact testing in air and in artificial seawater, as well as their corresponding 2D profiles. The radius and depth of wear craters are shown in Table 3.

Table 3. Sizes (radius, R, and depth, h) of wear craters of the coatings and steel substrate after impact testing in air and in ASW.

Material	In Air		In ASW	
	R, μm	h, μm	R, μm	h, μm
Substrate	400	28	430	31
0TiC	330	15	500	39
25TiC	300	13	310	12
50TiC	320	10	310	12
75TiC	270	10	290	8

The minimum radius and depth of wear craters were observed for the coating 75TiC with the maximum carbide content, both when tested in air (270 and 10 μm) and in ASW (290 and 8 μm). As the content of the carbide phase in the coatings decreased, an increase in the size of the wear craters was observed. For example, for a 50TiC and 25TiC coating, the radii and depth of wear craters in air were 320 and 10 μm, and 280 and 13 μm, respectively, while in ASW their values were identical and equaled to 310 and 12 μm. The maximum wear crater radii among coatings were observed for the 0TiC coating both in air (330 μm) and in sea water (500 μm). Interestingly, steel substrate possessed the lowest resistance to cyclic load in air, while the 0TiC coating performed the worst in ASW, despite the highest hardness. This discrepancy is due to the fact that despite the high hardness, the 0TiC coating is characterized by a very defective structure, which led to high brittleness and low fracture toughness in ASW experiments which are characterized by a more severe action than in air (link to our article.). The coatings with TiC were characterized by lower hardness values, but also by a high-quality defect-free structure which provided superior performance in impact testing both in air and in ASW.

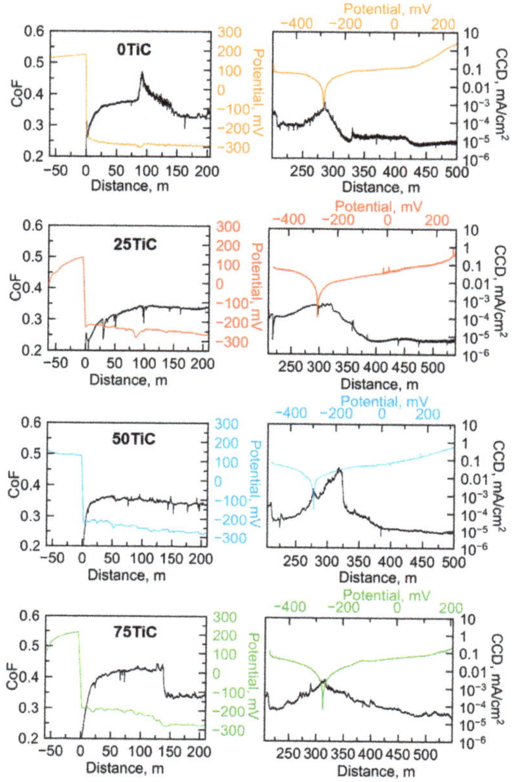

Figure 11. CoF and CCD vs. distance curves of the coatings tested against Al_2O_3 in ASW during combined tribocorrosion and electrochemical experiments.

The results of corrosion, tribocorrosion and impact tests in ASW are summarized in the scheme shown in Figure 13. The corrosion behavior of the TiC-free coating is determined by the formation of a Cr_2O_3-based passive film after the initial dissolution of iron (Figure 13a). High Cr content promotes dense and high-quality oxide layer formation.

Introduction of carbides into the coating (Figure 13b) leads to an increase in corrosion resistance via two mechanisms, which leads to more dense and thick passive film formation. First, TiC grains replace part of the surface, exposing a smaller area of the matrix. Second, TiC contributes to the enhanced passivation of the metal matrix because it has a more positive potential; however, this does not reach the repassivation potential of the metal matrix.

The protective properties of the coating under friction conditions are determined by the wear resistance of the solid phase and the rate of recovery of the passive film after its periodic removal by the counterbody and abrasive wear debris. The coating without carbide phase exhibits catastrophic wear both under friction and impact load due to high brittleness and low wear resistance. Additionally, under constant wear, passive film does not have time to recover completely between cycles (Figure 13c). A slight improvement in comparison with steel substrate is due to higher hardness and a higher content of corrosion-resistant components.

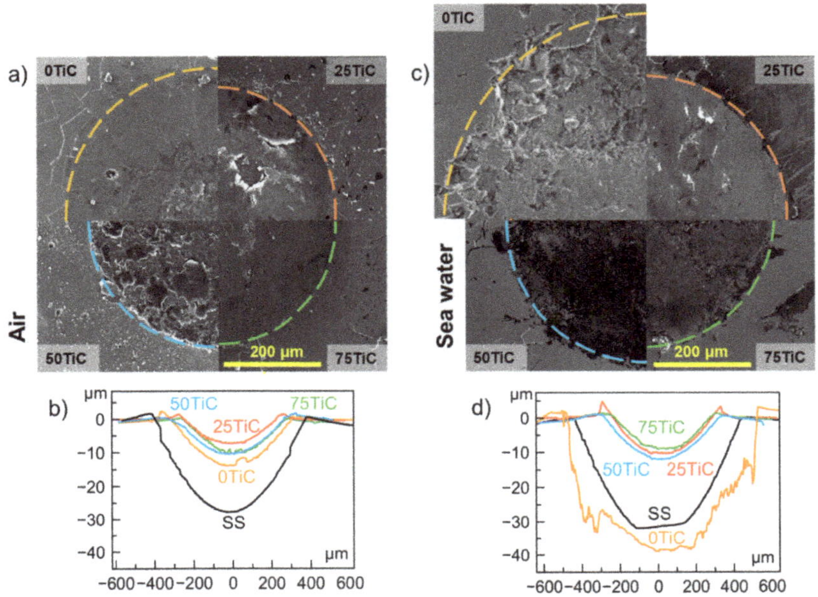

Figure 12. SEM images (**a**,**c**) and profiles of wear craters (**b**,**d**) of the coatings and steel substrate after impact testing in air and in ASW.

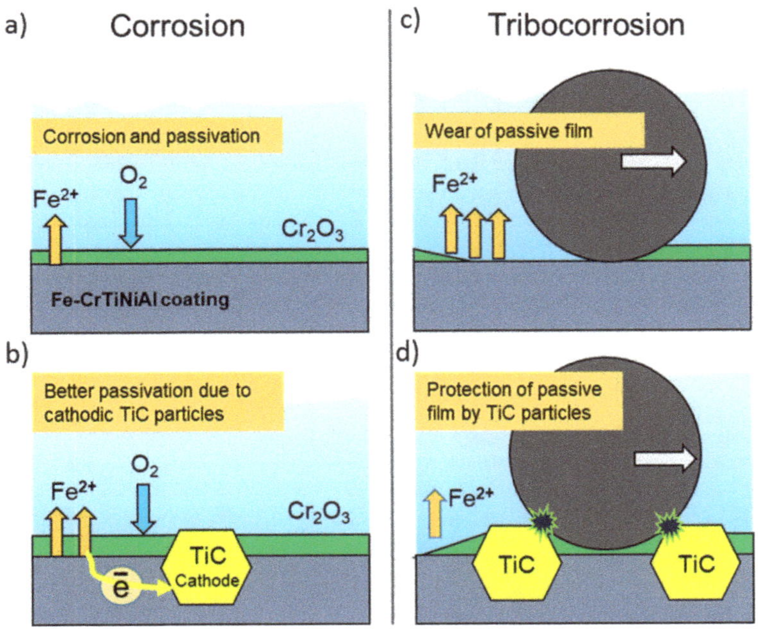

Figure 13. Scheme of TiC particles' effect on the corrosion behavior of the coatings under stationary (**a**,**c**) and wear conditions (**b**,**d**).

The best tribocorrosion behavior is demonstrated by coatings with high TiC content (Figure 13d). Under friction conditions, TiC grains, after the running-in stage, form reinforcing carcasses mainly contacting the counterbody and thereby protecting the passive film on the metal matrix from micro-cutting [21]. The positive electrochemical potential of carbides contributes to a more efficient passive film recovery on the worn areas of metal matrix. However, an excess of the carbide component is also undesirable, since it increases the abrasiveness of wear products.

4. Conclusions

In this work, novel composite xTiC-Fe-CrTiNiAl coatings with enhanced wear and corrosion resistance were successfully produced by the original technology of electrospark deposition in a vacuum. The effect of different TiC contents on structure, mechanical properties, corrosion and tribocorrosion behavior as well as cyclic impact loads was studied in detail. The results obtained may be summarized as follows:

(1) The TiC-free Fe-CrTiNiAl coating was characterized by a defective inhomogeneous structure with a large number of transverse and longitudinal cracks and a low Fe content of 20 at.%. The introduction of carbides changed the mass transfer mechanism during electrospark deposition; namely, Fe from the substrate intermixed with the electrode material more intensively (45–58 at.%) and led to the formation of coatings with a dense uniform structure, without visible defects and with evenly distributed TiC grains.

(2) With an increase in the carbide content in electrodes, the average size and content of TiC grains in the coatings increased. The TiC-containing coatings were characterized by a hardness of 10.3 GPa and an elastic modulus of 158 GPa. The formation of a composite structure with a heavily alloyed corrosion-resistant matrix based on an α-(Fe,Cr) solid solution and uniformly distributed TiC grains (100–200 nm in size) made it possible to significantly increase resistance to stationary corrosion and tribocorrosion in artificial seawater.

(3) The TiC-free coating demonstrated increased corrosion behavior compared to AISI 420S steel, possessing j_{corr} values of 0.2 and 9 $\mu A/cm^2$ under stationary and wear conditions compared to those of steel substrate (0.41 and 22 $\mu A/cm^2$); however, no enhancement of wear resistance was observed.

(4) Introduction of TiC led to a significant increase both in corrosion and tribocorrosion behavior. The best 75TiC-Cr_2Ti-NiAl coating demonstrated the lowest corrosion current densities both under stationary conditions (0.03 $\mu A/cm^2$) and under friction conditions (0.8 $\mu A/cm^2$). In addition, this coating was characterized by the lowest wear rate (4×10^{-6} mm^3/Nm vs. 1×10^{5} mm^3/Nm of TiC-free coating), a high resistance to cyclic impact loads (500 N, 10^5 cycles), the lowest potential shift during tribocorrosion (down to −240 mV compared to −340 mV for 25TiC-Fe-CrTiNiAl) and the fastest rate of recovery of the passive film under friction conditions.

(5) The mechanism of the enhancement of the corrosion and tribocorrosion behavior of the TiC-containing coatings was explained. Under stationary conditions, corrosion resistance is determined by high Cr content in the metal matrix and by TiC-induced passivation; meanwhile, under wear conditions, TiC particles acted as a reinforcing carcass, protecting the passive layer on metal matrix and promoting effective passivation of worn areas.

Author Contributions: Conceptualization, K.A.K. and A.N.S.; investigation, K.A.K., A.N.S., M.N.A. and A.V.B.; writing—original draft, M.N.A. and K.A.K.; writing—review and editing, D.V.S. and K.A.K.; visualization, K.A.K. and M.N.A.; supervision, K.A.K. and D.V.S.; project administration, K.A.K.; funding acquisition, K.A.K. All authors have read and agreed to the published version of the manuscript.

Funding: This work was financially supported by the Russian Science Foundation (Agreement No. 20-79-10104).

Institutional Review Board Statement: Not applicable.

Informed Consent Statement: Not applicable.

Data Availability Statement: Not applicable.

Acknowledgments: AVB acknowledges CzechNanoLab Research Infrastructure, supported by MEYS CR (LM2018110) for TEM investigations.

Conflicts of Interest: The authors declare no conflict of interest.

Abbreviations

CCD	Corrosion current density
OCP	Open-circuit corrosion potential
CoF	Coefficient of friction
ESD	Electrospark deposition
ASW	Artificial seawater
XRD	X-ray diffraction
SEM	Scanning electron microscopy
TEM	Transmitting electron microscopy
HAADF	High-angle annular dark field
EDS	Energy dispersive X-ray spectroscopy
SAED	Selected area electron diffraction
HR	TEM high resolution transmitting electron microscopy
DF	Dark field

References

1. Castro, J.D.; Lima, M.J.; Carvalho, S. Corrosion Resistance of Cu-Zr(O) N Films in a Simulated Seawater Environment. *Surf. Coat. Technol.* **2022**, *451*, 129050. [CrossRef]
2. Liu, W.; Yang, H.; Li, X.; Zhang, Z.; Lin, Y.; Deng, K. Effect of Chloride and Iodide on the Corrosion Behavior of 13Cr Stainless Steel. *Metals* **2022**, *12*, 1833. [CrossRef]
3. Olsson, C.O.A.; Landolt, D. Passive Films on Stainless Steels—Chemistry, Structure and Growth. *Electrochim. Acta* **2003**, *48*, 1093–1104. [CrossRef]
4. Wood, R.J.K. Marine Wear and Tribocorrosion. *Wear* **2017**, *376–377*, 893–910. [CrossRef]
5. Hoque, M.A.; Yao, C.W.; Khanal, M.; Lian, I. Tribocorrosion Behavior of Micro/Nanoscale Surface Coatings. *Sensors* **2022**, *22*, 9974. [CrossRef]
6. Rocha, A.M.F.; Bastos, A.C.; Cardoso, J.P.; Rodrigues, F.; Fernandes, C.M.; Soares, E.; Sacramento, J.; Senos, A.M.R.; Ferreira, M.G.S. Corrosion Behaviour of WC Hardmetals with Nickel-Based Binders. *Corros. Sci.* **2019**, *147*, 384–393. [CrossRef]
7. Alizadeh, M.; Mohammadi, G. Effect of Micro-Alloying Chromium on the Corrosion Resistance of Nanocrystalline Nickel Aluminide Intermetallic Produced by Mechanical Alloying Process. *Mater. Lett.* **2012**, *67*, 148–150. [CrossRef]
8. Cihangir, S.; Say, Y.; Ozkul, İ.; Guler, O.; Guler, S.H. Microstructure and Corrosion Investigation of $FeCoCrNiMo_{0,5}(MnAl)_{0,3}$ High Entropy Alloy Produced by 316 L Stainless Steel Scrap. *Mater. Today Commun.* **2022**, *33*, 104360. [CrossRef]
9. Lin, C.; Wu, W.; Han, Y.; Liu, J.; Zhang, M.; Wang, Q.; Li, X. Orderly Nucleation and Competitive Growth Behaviors of Ti-Al Intermetallic Compounds in Ti/TiAl3 Diffusion Couple under High Temperature. *J. Alloy. Compd.* **2023**, *939*, 168815. [CrossRef]
10. Singh, A.P.; Srivastava, C. Understanding the Non-Monotonic Variation in the Corrosion Rate of Sn-Ni Coatings with Ni Addition by the Analysis of Texture and Grain Boundary Constitution of the Matrix Phase and Spatial Distribution of the Intermetallic Phase in the Coating Microstructure. *Corros. Sci.* **2023**, *211*, 110787. [CrossRef]
11. Cai, C.; Song, B.; Wei, Q.; Xue, P.; Wen, S.; Liu, J.; Shi, Y. In-Situ Integrated Fabrication of Ti–Ni Coating during Hot Isostatic Pressing of Ti6Al4V Parts: Microstructure and Tribological Behavior. *Surf. Coat. Technol.* **2015**, *280*, 194–200. [CrossRef]
12. Chen, Y.; Niu, H.; Kong, F.; Xiao, S. Microstructure and Fracture Toughness of a β Phase Containing TiAl Alloy. *Intermetallics* **2011**, *19*, 1405–1410. [CrossRef]
13. Yokoshima, S.; Yamaguchi, M. Fracture Behavior and Toughness of PST Crystals of TiAl. *Acta Mater.* **1996**, *44*, 873–883. [CrossRef]
14. Ahledel, N.; Schulz, R.; Gariepy, M.; Hermawan, H.; Alamdari, H. Electrochemical Corrosion Behavior of Fe3Al/TiC and Fe3Al-Cr/TiC Coatings Prepared by HVOF in NaCl Solution. *Metals* **2019**, *9*, 437. [CrossRef]
15. Bin Humam, S.; Gyawali, G.; Amanov, A.; Kim, T.H.; Lee, S.W. Microstructure, Interface, and Nanostructured Surface Modifications to Improve Mechanical and Tribological Performance of Electrodeposited Ni-W-TaC Composite Coating. *Surf. Coat. Technol.* **2021**, *419*, 127293. [CrossRef]
16. Wang, Q.; Luo, S.; Wang, S.; Wang, H.; Ramachandran, C.S. Wear, Erosion and Corrosion Resistance of HVOF-Sprayed WC and Cr3C2 Based Coatings for Electrolytic Hard Chrome Replacement. *Int. J. Refract. Met. Hard Mater.* **2019**, *81*, 242–252. [CrossRef]
17. Santaella-González, J.B.; Hernández-Torres, J.; Morales-Hernández, J.; Flores-Ramírez, N.; Ferreira-Palma, C.; Rodríguez-Jiménez, R.C.; García-González, L. Effect of the Number of Bilayers in Ti/TiN Coatings on AISI 316L Deposited by Sputtering on Their Hardness, Adhesion, and Wear. *Mater. Lett.* **2022**, *316*, 132037. [CrossRef]
18. Fang, J.; Piliptsou, D.G.; Bekarevich, R.; Rogachev, A.V.; Jiang, X.; Kulesh, E. Effect of the Alloying Elements in TiN Sublayer on the Structure and Mechanical Properties of Carbon Coatings. *Thin Solid Films* **2022**, *755*, 139324. [CrossRef]
19. Ou, Y.X.; Wang, H.Q.; Hua, Q.S.; Liao, B.; Ouyang, X.P. Tribocorrosion Behaviors of Superhard yet Tough Ti-C-N Ceramic Coatings. *Surf. Coat. Technol.* **2022**, *439*, 128448. [CrossRef]

20. Zou, Y.; Tan, C.; Qiu, Z.; Ma, W.; Kuang, M.; Zeng, D. Additively Manufactured SiC-Reinforced Stainless Steel with Excellent Strength and Wear Resistance. *Addit. Manuf.* **2021**, *41*, 101971. [CrossRef]
21. Kan, W.H.; Proust, G.; Bhatia, V.; Chang, L.; Dolman, K.; Lucey, T.; Tang, X.; Cairney, J. Slurry Erosion, Sliding Wear and Corrosion Behavior of Martensitic Stainless Steel Composites Reinforced in-Situ with NbC Particles. *Wear* **2019**, *420–421*, 149–162. [CrossRef]
22. Wang, X.; Wang, M.; Yi, H.; García, J.L.; Wang, X.; Wang, M. Research on the Preparation Process and Performance of a Wear-Resistant and Corrosion-Resistant Coating. *Crystals* **2022**, *12*, 591. [CrossRef]
23. Antonyuk, M.N.; Kuptsov, K.A.; Sheveyko, A.N.; Shtansky, D.V. Antibacterial TaC-(Fe,Cr,Mo,Ni)-(Ag/Cu) Composite Coatings with High Wear and Corrosion Resistance in Artificial Seawater. *Lubricants* **2022**, *10*, 320. [CrossRef]
24. Wu, T.; Chen, Y.; Lin, B.; Yu, L.; Gui, W.; Li, J.; Wu, Y.; Zeng, D. Effects of WC on the Microstructure, Wear and Corrosion Resistance of Laser-Deposited CoCrFeNi High Entropy Alloy Coatings. *Coatings* **2022**, *12*, 985. [CrossRef]
25. Li, W.; Li, J.; Xu, Y. Investigation into the Corrosion Wear Resistance of CoCrFeNiAlx Laser-Clad Coatings Mixed with the Substrate. *Metals* **2022**, *12*, 460. [CrossRef]
26. Tang, M.; Zhang, L.; Zhang, N. Microstructural Evolution, Mechanical and Tribological Properties of TiC/Ti6Al4V Composites with Unique Microstructure Prepared by SLM. *Mater. Sci. Eng. A* **2021**, *814*, 141187. [CrossRef]
27. Günen, A.; Soylu, B.; Karakaş, Ö. Titanium Carbide Coating to Improve Surface Characteristic, Wear and Corrosion Resistance of Spheroidal Graphite Cast Irons. *Surf. Coat. Technol.* **2022**, *437*, 128280. [CrossRef]
28. Kayalı, Y.; Kanca, E.; Günen, A. Effect of Boronizing on Microstructure, High-Temperature Wear and Corrosion Behavior of Additive Manufactured Inconel 718. *Mater. Charact.* **2022**, *191*, 112155. [CrossRef]
29. Siddiqui, A.R.; Maurya, R.; Katiyar, P.K.; Balani, K. Superhydrophobic, Self-Cleaning Carbon Nanofiber CVD Coating for Corrosion Protection of AISI 1020 Steel and AZ31 Magnesium Alloys. *Surf. Coat. Technol.* **2020**, *404*, 126421. [CrossRef]
30. Liao, L.; Gao, R.; Yang, Z.H.; Wu, S.T.; Wan, Q. A Study on the Wear and Corrosion Resistance of High-Entropy Alloy Treated with Laser Shock Peening and PVD Coating. *Surf. Coat. Technol.* **2022**, *437*, 128281. [CrossRef]
31. Feng, J.; Xiao, H. Tribocorrosion Behavior of Laser Cladded Ti-Al-(C, N) Composite Coatings in Artificial Seawater. *Coatings* **2022**, *12*, 187. [CrossRef]
32. Aubry, P.; Blanc, C.; Demirci, I.; Gorny, C.; Maskrot, H. Analysis of a Ni-Fe-Cr-Mo-Si Hardfacing Alloy Manufactured by Laser Cladding: Influence of the Iron Content on the Wear Resistance Properties. *Procedia CIRP* **2018**, *74*, 210–213. [CrossRef]
33. Brezinová, J.; Landová, M.; Guzanová, A.; Dulebová, L.; Draganovská, D. Microstructure, Wear Behavior and Corrosion Resistance of WC-FeCrAl and WC-WB-Co Coatings. *Metals* **2018**, *8*, 399. [CrossRef]
34. Klekotka, M.; Zielińska, K.; Stankiewicz, A.; Kuciej, M. Tribological and Anticorrosion Performance of Electroplated Zinc Based Nanocomposite Coatings. *Coatings* **2020**, *10*, 594. [CrossRef]
35. Kuptsov, K.A.; Kiryukhantsev-Korneev, P.V.; Sheveyko, A.N.; Shtansky, D.V. Comparative Study of Electrochemical and Impact Wear Behavior of TiCN, TiSiCN, TiCrSiCN, and TiAlSiCN Coatings. *Surf. Coat. Technol.* **2013**, *216*, 273–281. [CrossRef]
36. Kuptsov, K.A.; Sheveyko, A.N.; Manakova, O.S.; Sidorenko, D.A.; Shtansky, D.V. Comparative Investigation of Single-Layer and Multilayer Nb-Doped TiC Coatings Deposited by Pulsed Vacuum Deposition Techniques. *Surf. Coat. Technol.* **2020**, *385*, 125422. [CrossRef]
37. Kuptsov, K.A.; Sheveyko, A.N.; Zamulaeva, E.I.; Sidorenko, D.A.; Shtansky, D.V. Two-Layer Nanocomposite WC/a-C Coatings Produced by a Combination of Pulsed Arc Evaporation and Electro-Spark Deposition in Vacuum. *Mater. Des.* **2019**, *167*, 107645. [CrossRef]
38. Kuptsov, K.A.; Sheveyko, A.N.; Sidorenko, D.A.; Shtansky, D.V. Electro-Spark Deposition in Vacuum Using Graphite Electrode at Different Electrode Polarities: Peculiarities of Microstructure, Electrochemical and Tribological Properties. *Appl. Surf. Sci.* **2021**, *566*, 150722. [CrossRef]
39. Pezzato, L.; Vranescu, D.; Sinico, M.; Gennari, C.; Settimi, A.G.; Pranovi, P.; Brunelli, K.; Dabalà, M. Tribocorrosion Properties of PEO Coatings Produced on AZ91 Magnesium Alloy with Silicate- or Phosphate-Based Electrolytes. *Coatings* **2018**, *8*, 202. [CrossRef]
40. Petrov, Y.I.; Shafranovsky, E.A.; Krupyanskii, Y.F.; Essine, S.V. Structure and Mössbauer Spectra for the Fe–Cr System: From Bulk Alloy to Nanoparticles. *J. Appl. Phys.* **2001**, *91*, 352. [CrossRef]
41. Tornabene, F.; Lu, C.-J.; Yeh, J.-W. Improved Wear and Corrosion Resistance in TiC-Reinforced SUS304 Stainless Steel. *J. Compos. Sci.* **2023**, *7*, 34. [CrossRef]
42. Sheveyko, A.N.; Kuptsov, K.A.; Kiryukhantsev-Korneev, P.V.; Kaplansky, Y.Y.; Orekhov, A.S.; Levashov, E.A. Protective Coatings for LPBF Ni-Based Superalloys Using a Combination of Electrospark Deposition and Pulsed Arc Evaporation Methods. *Appl. Surf. Sci.* **2022**, *581*, 152357. [CrossRef]
43. Zhang, J.; Xin, S.; Zhang, Y.; Guo, P.; Sun, H.; Li, T.; Qin, C. Effects of Elements on the Microstructure and Mechanical Properties of AlCoCrFeNiTi High-Entropy Alloys. *Metals* **2023**, *13*, 178. [CrossRef]
44. Memarrashidi, Z.; Plucknett, K.P. The Effects of C:N Ratio on the Aqueous Corrosion Response of TiC and Ti(C,N) Cermets with a Ni3Al-Based Binder. *Int. J. Refract. Met. Hard Mater.* **2016**, *61*, 162–172. [CrossRef]

Disclaimer/Publisher's Note: The statements, opinions and data contained in all publications are solely those of the individual author(s) and contributor(s) and not of MDPI and/or the editor(s). MDPI and/or the editor(s) disclaim responsibility for any injury to people or property resulting from any ideas, methods, instructions or products referred to in the content.

coatings

MDPI

Article

CMAS Corrosion Resistance Behavior and Mechanism of $Hf_6Ta_2O_{17}$ Ceramic as Potential Material for Thermal Barrier Coatings

Sai Liu [1,†], Qing Liu [1,†], Xiaopeng Hu [1], Jinwei Guo [1], Wang Zhu [1,*], Fan Zhang [1,*] and Jie Xia [2]

1 Key Laboratory of Key Film Materials & Application for Equipment (Hunan Province), School of Materials Science and Engineering, Xiangtan University, Xiangtan 411105, China
2 Department of Mechanical Engineering, Hunan Institute of Technology, Hengyang 421002, China
* Correspondence: wzhu@xtu.edu.cn (W.Z.); zhangfan15@xtu.edu.cn (F.Z.)
† These authors contributed equally to this work.

Abstract: Thermal barrier coatings (TBCs) have been seriously threatened by calcium-magnesium-alumina-silicate (CMAS) corrosion. The search for novel ceramic coatings for TBCs with excellent resistance to CMAS corrosion is ongoing. Herein, CMAS corrosion resistance behavior and the mechanism of a promising $Hf_6Ta_2O_{17}$ ceramic coating for TBCs are investigated. The results show that temperature is the most important factor affecting the CMAS behavior and mechanism. At 1250 °C, the corrosion products are composed of dense reaction products ($HfSiO_4$, $Ca_XHf_{6-x}Ta_2O_{17-x}$) and CMAS self-crystallization products. At 1300 and 1400 °C, the corrosion products are mainly dense $CaTa_2O_6$ and HfO_2, which prevent further CMAS infiltration.

Keywords: thermal barrier coatings (TBCs); calcium-magnesium-alumina-silicate (CMAS); $Hf_6Ta_2O_{17}$ ceramic; corrosion behavior; corrosion resistance mechanism

1. Introduction

Thermal barrier coatings (TBCs) can effectively protect hot components in high temperature environments, and improve gas thermal efficiency and engine service lifetime [1–3]. 7–8 wt.% Y_2O_3 partially stabilized ZrO_2 (8YSZ) is extensively used as a top ceramic coating in TBCs. However, erosion, sintering, oxidation and calcium-magnesium-alumina-silicate (CMAS) corrosion are the main failure factors of TBCs in high- temperature service environments. As the service temperature increases, CMAS corrosion becomes the most dangerous factor in spallation of the top ceramic coating [4–6]. On the one hand, the molten CMAS permeates into the pores and microcracks of the 8YSZ coating, leading to the degradation of the thermal-mechanical properties of the coating [7,8]. On the other hand, Y_2O_3 in 8YSZ dissolves in the molten CMAS through thermal-chemical reaction, which induces a phase transition of 8YSZ. This destroys the structural integrity and leads to coating failure [9,10]. Hence, it is significant to seek novel ceramic coatings for TBCs with excellent resistance to CMAS corrosion.

Recently, many studies have been conducted on the promising CMAS-resistant TBCs, such as $RE_2Zr_2O_7$, $RETaO_4$, $REPO_4$ (RE = rare earth elements), Ti_2AlC, $LaTi_2Al_9O_{19}$ and rare-earth-doped zirconia [11–16]. These new ceramic materials show different corrosion behaviors and mechanisms. For $LaTi_2Al_9O_{19}$ ceramics, Ca and Si elements in CMAS are consumed effectively to form $CaAl_2Si_2O_8$ and $(Ca,La_{0.7})(Ti,Al)O_3$, which prevent CMAS from further infiltration [14]. The corrosion mechanism the changing of the composition of molten CMAS and generating self-crystallization products that prevent the CMAS infiltration. For $RETaO_4$ and $RE_2Zr_2O_7$ ceramics, $YbTaO_4$ reacts with CMAS to form $Ca_2Ta_2O_7$ [12], and $La_2(Zr_{0.7}Ce_{0.3})_2O_7$ reacts with CMAS to precipitate apatite phase $(La,Ca)_4(La, Ce)_6(SiO_4)_6O_2$ and fluorite ZrO_2 [17]. The corrosion mechanism is that the ceramic material reacts with molten CMAS to form a dense product layer, inhibiting further CMAS infiltration. These new ceramic materials possess outstanding CMAS corrosion resistance. However, they still cannot replace the traditional 8YSZ TBCs due to their mismatched thermal expansion coefficients [18] and insufficient comprehensive performance of thermal and mechanical properties [19,20]. $Hf_6Ta_2O_{17}$ ceramic material with $A_6B_2O_{17}$ (A = Hf, Zr, B = Ta, Nb) superstructure crystal can solve the above problems effectively [21,22]. $Hf_6Ta_2O_{17}$ ceramic has a sufficiently large synthesis range, superior phase stability and excellent thermal properties [23–25]. Furthermore, $Hf_6Ta_2O_{17}$ ceramic shows excellent mechanical properties, with hardness of 18.45 GPa and fracture toughness of 2.6~3.1 MPa $m^{1/2}$ at room temperature [26]. The comprehensive properties of $Hf_6Ta_2O_{17}$ ceramic are excellent, but there is a lack of systematic in-depth study on the CMAS corrosion resistance behavior and mechanism of a potential $Hf_6Ta_2O_{17}$ ceramic for TBCs.

In this paper, $Hf_6Ta_2O_{17}$ ceramic is prepared by a solid reaction method, and the $Hf_6Ta_2O_{17}$ ceramic is subjected to CMAS corrosion at 1250, 1300 and 1400 °C for 4, 16, 50 and 100 h each. The corrosion behavior is investigated and the effects of temperature and time on CMAS corrosion resistance of the $Hf_6Ta_2O_{17}$ ceramic are discussed. The research plan is shown in Figure 1.

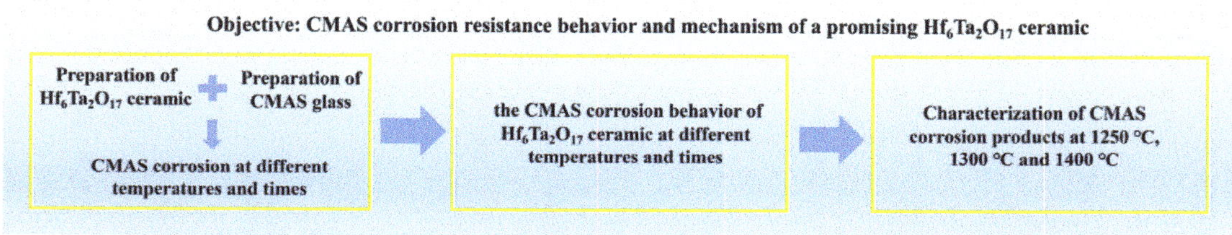

Figure 1. Schematic flowchart showing research plan.

2. Experimental Procedure

HfO_2 (purity > 99.99%) and Ta_2O_5 (purity > 99.99%) were selected as the starting materials (both oxides were purchased from China New Metal Materials Technology Co., Ltd., Bejing, China). The powders were weighted in a stoichiometric ratio of 85.71 mol% HfO_2–14.29 mol% Ta_2O_5 and then mixed with ethanol by a planetary ball mill for 8 h. After drying the slurries at 80 °C for 12 h, the mixed powders were reacted at 1600 °C for 20 h to obtain $Hf_6Ta_2O_{17}$ powder. The chemical reaction equation is presented as follows:

$$6HfO_2 + Ta_2O_5 \rightarrow Hf_6Ta_2O_{17} \quad (1)$$

Finally, the $Hf_6Ta_2O_{17}$ powders were sieved with 500-mesh screen, and subsequently uniaxially cold compacted into pellets (φ = 10 mm) under a pressure of 200 MPa. The $Hf_6Ta_2O_{17}$ ceramic was obtained by sintering the pellets at 1600 °C for 20 h. Based on CMAS coated on vane blades in a real service environment, the CMAS composition of 33 mol% CaO–9 mol% MgO–13 mol%–Al_2O_3–45 mol% SiO_2 was determined, whose melting point is about 1235 °C [27,28]. The oxide powders were weighed and mixed in the planetary ball mill in these proportions. After drying, CMAS glass was synthesized by heat treating at 1300 °C for 4 h, then ground to get fine CMAS powder.

Before the CMAS corrosion test, the $Hf_6Ta_2O_{17}$ ceramic was ground, polished and ultrasonically cleaned. According to the surface area of the prepared ceramic pellets, an appropriate amount of CMAS powder was weighed with an analytical balance. To evenly coat the $Hf_6Ta_2O_{17}$ ceramic surface, the CMAS powder was mixed with absolute alcohol to create a pasty, viscous liquid. After drying at 80 °C, the coated CMAS amount was kept at 10 ± 1 mg/cm^2. Then, samples of the CMAS-covered ceramics were heat treated at 1250, 1300 and 1400 °C, with a ramping rate of 10 °C/min for both heating and cooling, in a muffle furnace for 4, 16, 50 and 100 h each. Three samples were put into the furnace for comparison in each test. Finally, the corroded ceramics were cut through the middle with a cutting machine to obtain the cross section.

The phase structures of the $Hf_6Ta_2O_{17}$ ceramic and corrosion products were detected by X-ray diffractometry (XRD, Ultimate IV, RIGAKU, Tokyo, Japan) with CuKα radiation at a scanning rate of 4°/min and scanning range of 10~90°. The microstructure and composition of the specimens were measured by scanning electron microscope (SEM, TESCAN MIRA3 LMH, Brno, Czech Republic), energy dispersive spectroscopy (EDS, X MAX20, Oxford Instruments, Oxford, UK) and transmission electron microscopy (TEM, JEM-2100, JEOL Ltd., Tokyo, Japan). Image J software (1.51 23) was used for image analysis, and the infiltration depth was measured by cross-sectional backscattered electron (BSE) image after corrosion. Five different cross-sectional BSE images were taken, with each image taking five depth measurements. Finally, the average value was taken to obtain the infiltration depth.

3. Results

3.1. Characterization of $Hf_6Ta_2O_{17}$ Ceramic

Figure 2 shows the XRD pattern and SEM image of the surface of $Hf_6Ta_2O_{17}$ ceramic. The characteristic peak is a pure structure of orthorhombic $Hf_6Ta_2O_{17}$ phase, and no diffraction peaks of HfO_2 or Ta_2O_5 phase are detected. The $Hf_6Ta_2O_{17}$ ceramic has a dense microstructure (with porosity below 15%), and no obvious compositional contrast appears, which implies that no possible component segregation occurs.

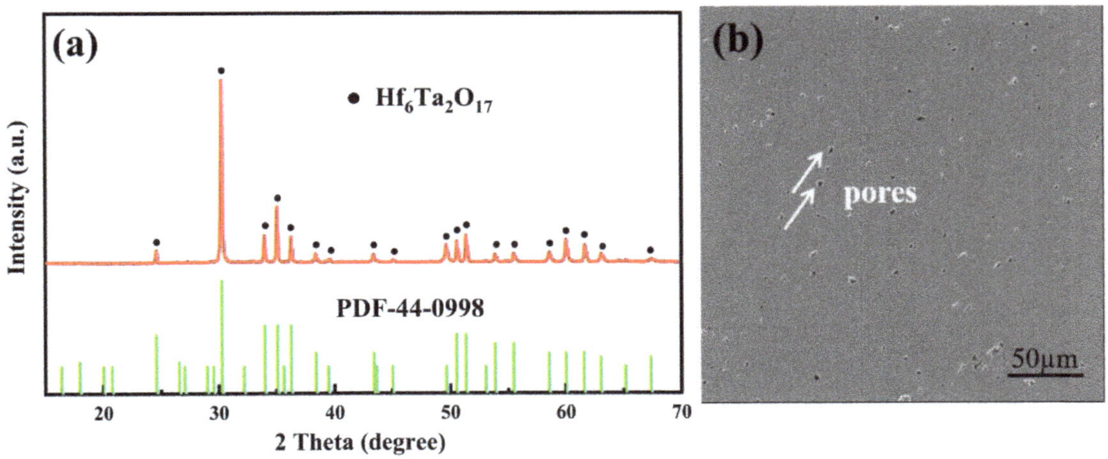

Figure 2. (**a**) XRD patterns of as-sintered $Hf_6Ta_2O_{17}$ ceramic; (**b**) surface morphology of $Hf_6Ta_2O_{17}$ ceramic.

3.2. Microstructure Characterization of $Hf_6Ta_2O_{17}$ Ceramic after CMAS Corrosion

3.2.1. Surface Morphology of $Hf_6Ta_2O_{17}$ Ceramic after CMAS Corrosion

The surface morphology of $Hf_6Ta_2O_{17}$ ceramic after CMAS corrosion is shown in Figure 3. At the corrosion temperature of 1250 °C, there is a black strip phase (A) and a gray bulk phase (B), in addition to residual CMAS melt (C). The elemental composition of the marked areas is listed in Table 1. A is anorthite phase ($CaAl_2Si_2O_8$) and B is wollastonite phase ($CaSiO_3$), both of which are self-crystallization products of molten CMAS [29]. As the corrosion temperature elevates to 1300 °C, the specimen surface is mainly composed of residual CMAS (C) and a little white snowflake phase (D) that is rich in Ca, Hf and Ta elements. The corrosion results at 1400 °C reveal that a large amount of white dendritic phase (E) appears on the specimen surface. The composition of the white dendritic phase (E) is similar to that of D, which is rich in Ca, Hf and Ta elements, but the Ta content is significantly higher than that of D. The corrosion results indicate that the corrosion products and corrosion phenomena of $Hf_6Ta_2O_{17}$ ceramic are obviously different at different corrosion temperatures.

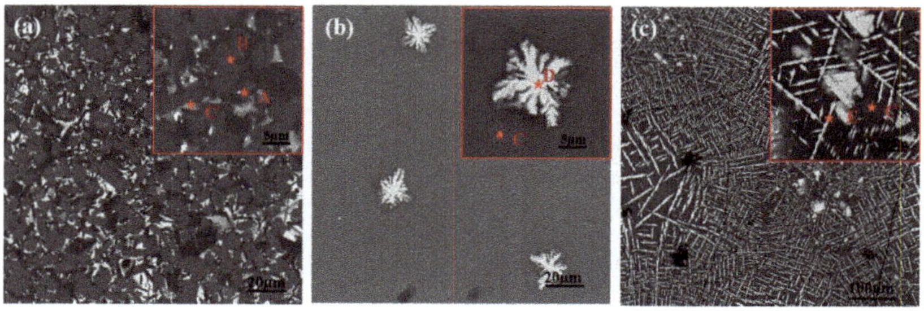

Figure 3. BSE images of the surface of $Hf_6Ta_2O_{17}$ ceramic after CMAS corrosion: (**a**) at 1250 °C for 8 h; (**b**) at 1300 °C for 16 h; (**c**) at 1400 °C for 16 h. The inset figures present high-magnification images of the surface.

Table 1. Chemical compositions of the marked regions in Figure 3.

Fraction Location	Composition (at.%)							
	O	Mg	Al	Si	Ca	Hf	Ta	-
A	58.34	0.13	16.09	17.23	8.17	0.02	0.02	$CaAl_2Si_2O_8$
B	56.16	0.18	1.50	21.55	19.94	0.46	0.21	$CaSiO_3$
C	56.40	0.9	1.7	21.6	17.2	1.6	0.6	CMAS
D	60.57	0.30	1.02	-	11.05	16.00	11.06	-
E	51.90	0.01	2.37	8.62	13.59	3.08	20.45	-

3.2.2. Cross-Sectional Morphology of $Hf_6Ta_2O_{17}$ Ceramic after CMAS Corrosion

Figure 4 shows the cross-sectional morphology of $Hf_6Ta_2O_{17}$ ceramic attacked by CMAS at different temperatures and times. The cross section is obviously stratified into a residual CMAS layer, reaction layer, dense layer and ceramic layer. The corrosion behavior of $Hf_6Ta_2O_{17}$ ceramic at three temperatures is dramatically distinctive. At 1250 °C, when the corrosion time is short (<100 h), the infiltration depth of molten CMAS is very shallow (Figure 4a–d), and the thickness of the reaction layer is in the range of 3.3~15.2 μm. After 100 h, an obvious dense layer with a thickness of 22.2 μm appears below the reaction layer (Figure 4d). At 1300 °C for 4 h, the reaction products are similar to those at 1250 °C, but a new phase structure begins to precipitate (Figure 4e), which is analyzed in detail in Section 3.3.2. When the corrosion time increases to 16 h, the morphology of the reaction product changes dramatically, and CMAS infiltration depth increases to 34.1 μm (Figure 4f). Furthermore, the thickness of the reaction layer increases slowly with time, ranging from 17.8 to 27.9 μm, but the dense layer keeps thickening, increasing from 16.3 to 125.1 μm (Figure 4f–h). The corrosion phenomenon at 1400 °C is similar to that at 1300 °C (Figure 4j–l). At the initial stage of corrosion, a thick reaction layer with a thickness of 83.3 μm is generated, while the dense layer is extremely thin (~20.6 μm). With the increase in corrosion time from 4 h to 100 h, the thickness of the reaction layer is basically stable, increasing by only about 10 μm. The dense layer also becomes thicker, increasing from 20.6 to 86.6 μm.

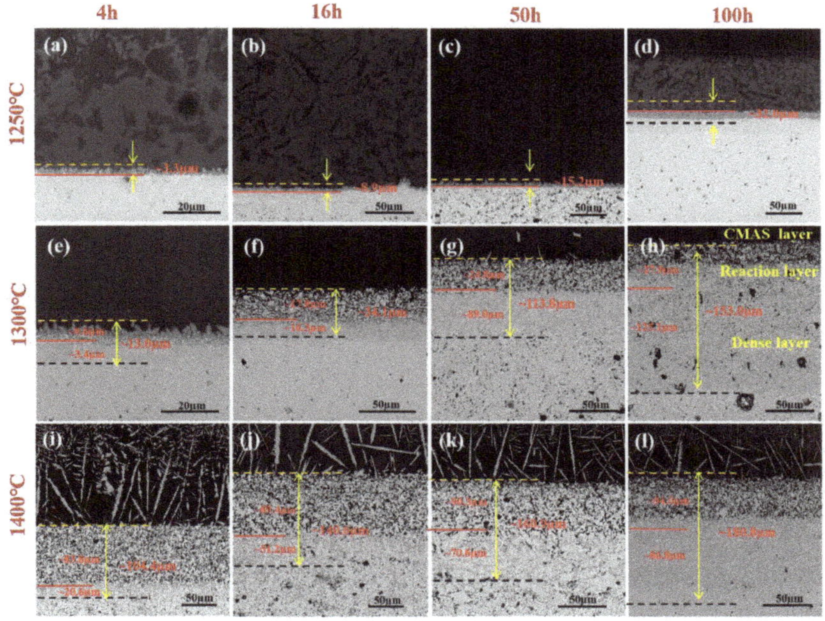

Figure 4. Cross-sectional BSE images of $Hf_6Ta_2O_{17}$ ceramic after CMAS corrosion: (**a**) at 1250 °C for 4 h; (**b**) at 1250 °C for 16 h; (**c**) at 1250 °C for 50 h; (**d**) at 1250 °C for 100 h; (**e**) at 1300 °C for 4 h; (**f**) at 1300 °C for 16 h; (**g**) at 1300 °C for 50 h; (**h**) at 1300 °C for 100 h; (**i**) at 1400 °C for 4 h; (**j**) at 1400 °C for 16 h; (**k**) at 1400 °C for 50 h; (**l**) at 1400 °C for 100 h. The CMAS infiltration depth is measured from the initial specimen surface (yellow dash lines) to the infiltration front (black dash lines).

The evolution of CMAS infiltration depth with corrosion time of $Hf_6Ta_2O_{17}$ ceramic at three temperatures is shown in Figure 5. After CMAS corrosion for 100 h, the CMAS infiltration depths at 1250, 1300 and 1400 °C are 32.3, 153 and 180.8 μm, respectively. $Hf_6Ta_2O_{17}$ ceramic shows a lower infiltration depth (32.3 μm) at 1250 °C. At 1300 and 1400 °C, the infiltration depth of $Hf_6Ta_2O_{17}$ ceramic increases significantly. After fitting the experimental data, the infiltration rate k at the three temperatures can be quantified in the form of a $\mu m \cdot t^n$, where a and n are constants and t is time. At 1250, 1300 and 1400 °C, the infiltration rates of $Hf_6Ta_2O_{17}$ ceramic are 0.85 $\mu m \cdot t^{-0.29}$, 3.6 $\mu m \cdot t^{-0.23}$ and 14.17 $\mu m \cdot t^{-0.83}$, respectively. The parameter a increases significantly with increasing temperature, and the difference between n at 1250 and 1300 °C is not significant, while a notable decrease occurs at 1400 °C. This indicates that the infiltration rate at 1300 °C is significantly greater than that at 1250 °C. The infiltration rate also appears to increase at 1400 °C, while the enhancement is much greater than that at 1300 °C. The CMAS infiltration depth reaches 104.4 μm after corrosion for 4 h at 1400 °C, indicating that temperature is the most sensitive factor influencing the CMAS corrosion behavior of $Hf_6Ta_2O_{17}$ ceramic. Furthermore, the infiltration depth increases with time, and the infiltration rate decreases with time. The relationship between CMAS infiltration depth and corrosion time basically follows a parabolic law, which is due to the formation and growth of the dense layer [30].

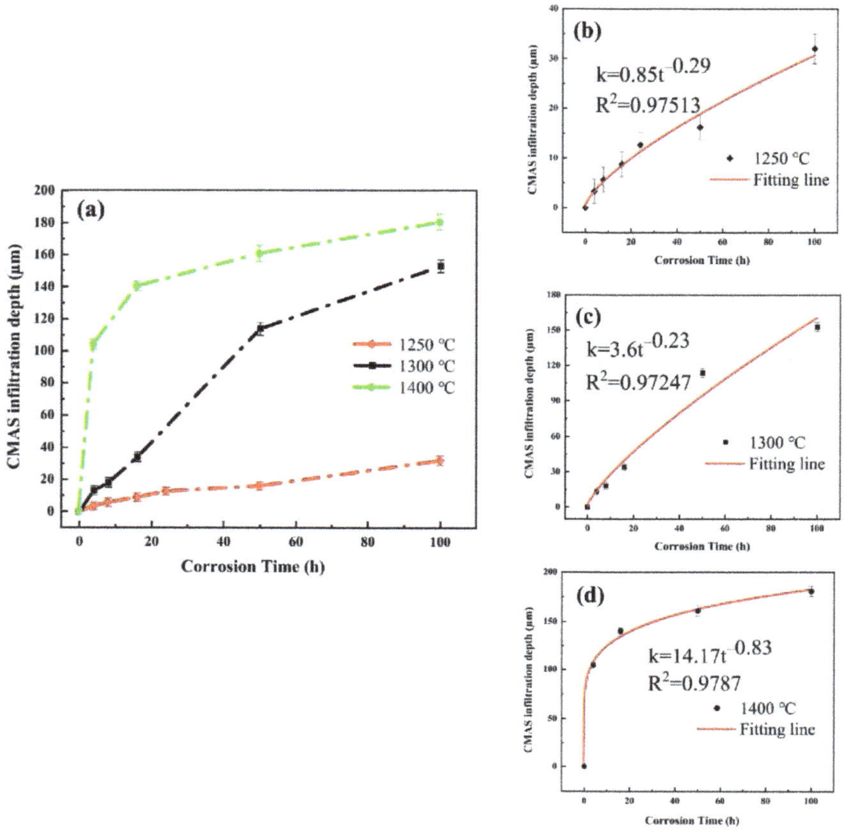

Figure 5. (**a**) Evolution of CMAS infiltration depth with time in $Hf_6Ta_2O_{17}$ ceramic at different temperatures; (**b–d**) Expanded view of infiltration depth with time in $Hf_6Ta_2O_{17}$ ceramic at 1250, 1300 and 1400 °C, respectively. R^2 represents the value of the matching factor.

3.3. Characterization of CMAS Corrosion Products of $Hf_6Ta_2O_{17}$ Ceramic

3.3.1. Characterization of Corrosion Products at 1250 °C

Figure 6 shows a cross-sectional image of $Hf_6Ta_2O_{17}$ ceramic attacked by CMAS at 1250 °C for 4 h and its corresponding element mapping. There are two main types of products in the residual CMAS layer: A and B. The reaction layer is only 2~3 μm thick, made up of irregular lumps. At the same time, Ca element infiltrates into the ceramic layer, forming a thin dense layer. The element mapping results show that the segregation of Ca and Al appears in the residual CMAS layer, a small amount of Hf diffuses into the residual CMAS, and the content of Ta is extremely low. This is attributed to the fact that the Hf-O bond in $Hf_6Ta_2O_{17}$ ceramic is longer than the Ta-O bond, which implies that Hf element more easily diffuses into the residual CMAS [21]. XRD patterns of $Hf_6Ta_2O_{17}$ ceramic attacked by CMAS at 1250 °C for 4, 16, 50 and 100 h are shown in Figure 7. The characteristic peaks of the corrosion products are mainly $CaAl_2Si_2O_8$ and $CaSiO_3$ after corrosion for 4 h. However, the $CaSiO_3$ phase disappears after corrosion for 16 h and the characteristic peak of $HfSiO_4$ phase appears with time. The highest diffraction peaks of anorthite are detected under long-term corrosion. According to the XRD results and the EDS results in Table 2, it can be determined that the molten CMAS self-crystallizes into $CaAl_2Si_2O_8$ (A) and $CaSiO_3$ (B), the reaction layer consists of $HfSiO_4$ (D), and the dense layer is composed of $Ca_XHf_{6-x}Ta_2O_{17-x}$ (F) after corrosion for 4 h at 1250 °C. It is worth noting that there are no characteristic peaks of $HfSiO_4$ phase after corrosion for a short time, because few $HfSiO_4$ phases are formed under short-term corrosion and the residual CMAS is quite thick.

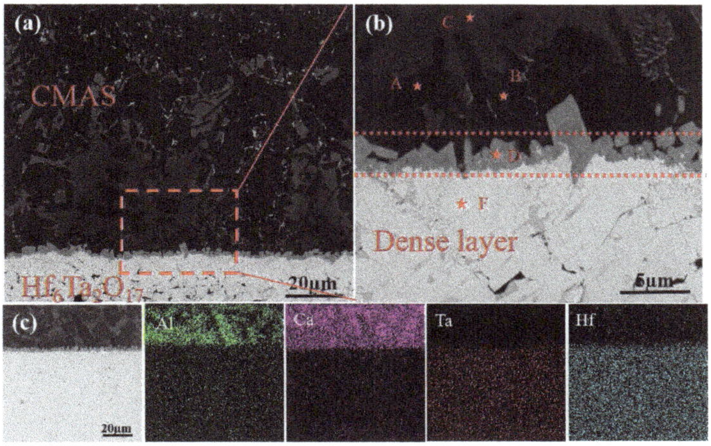

Figure 6. (**a**,**b**) Cross-sectional BSE images of $Hf_6Ta_2O_{17}$ ceramic after CMAS corrosion at 1250 °C for 4 h; (**c**) cross-sectional EDS mapping results of $Hf_6Ta_2O_{17}$ ceramic corroded by CMAS.

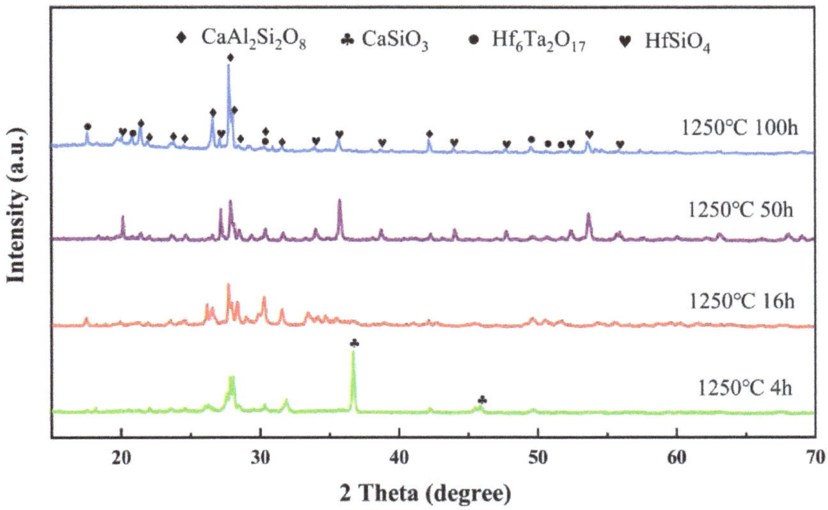

Figure 7. XRD patterns of $Hf_6Ta_2O_{17}$ ceramic corroded by CMAS at 1250 °C for 4 h, 16 h, 50 h and 100 h.

Table 2. Chemical compositions of the marked regions in Figure 6b.

Fraction Location	Composition (at.%)							
	O	Mg	Al	Si	Ca	Hf	Ta	-
A	58.15	0.36	14.29	17.92	8.89	0.25	0.14	$CaAl_2Si_2O_8$
B	56.88	0.09	1.34	20.61	20.63	0.14	0.10	$CaSiO_3$
C	56.50	1.37	5.41	20.74	13.13	2.00	0.85	CMAS
D	62.20	0.18	2.43	13.74	11.27	8.76	1.43	$HfSiO_4$
F	68.20	-	0.51	-	4.84	20.52	5.93	$Ca_XHf_{6-x}Ta_2O_{17-x}$

Figure 8 shows a cross-sectional image of $Hf_6Ta_2O_{17}$ ceramic corroded by CMAS at 1250 °C for 16 h and its corresponding EDS line scan. It can be seen that there is only anorthite phase ($CaAl_2Si_2O_8$) in the residual CMAS. The wollastonite phase ($CaSiO_3$) disappears at 16 h, corresponding to the XRD result (Figure 7). To further characterize the dense layer, an EDS line scan is performed on the reactive layer and the dense layer in Figure 8a. The content of Ca element in the dense layer decreases in a gradient manner with the increase in the depth, as shown in Figure 8b, which proves that the Ca infiltration in $Hf_6Ta_2O_{17}$ ceramic is the result of volume diffusion rather than grain boundary diffusion. In addition, the Hf element fluctuates in the dense layer, as shown in the dashed circle in Figure 8b, which may be due to the generation of HfO_2 phase or Hf element segregation in the dense layer. To further determine the composition, TEM is used to characterize the interface between the reaction layer and dense layer (see Figure 9). Figure 9a,b are the bright-field (BF) TEM images of CMAS reaction front corresponding to the dashed rectangle in Figure 8a. It can be seen that there is a distinct interface between the reaction layer and the dense layer. According to the selected area electron diffraction (SAED) presented in Figure 9d,e, it can be determined that the product D is $HfSiO_4$ and E is the orthorhombic $Hf_6Ta_2O_{17}$ phase. Furthermore, no other phase structure is found in the dense layer, which confirms that m-HfO_2 is not generated and only Hf element segregation exists. The TEM EDS mapping between the reactive layer and the dense layer in Figure 9c again proves that Ca element penetrates into the dense layer and dose not penetrate down along the grain boundary.

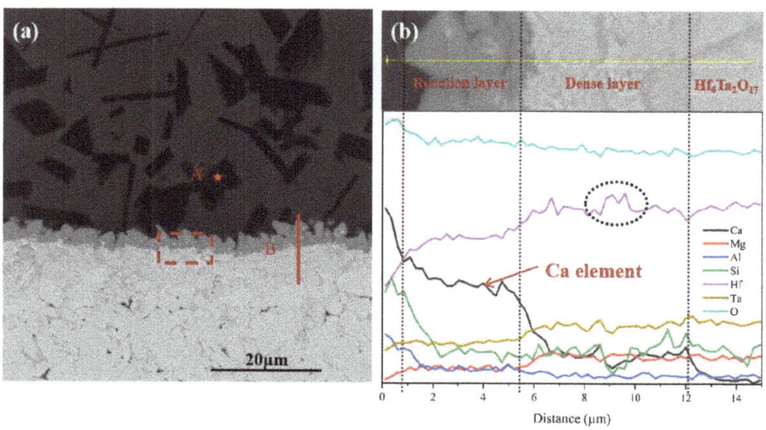

Figure 8. (**a**) Cross-sectional BSE images of $Hf_6Ta_2O_{17}$ after CMAS attack at 1250 °C for 16 h; (**b**) EDS line scan in location B of (**a**).

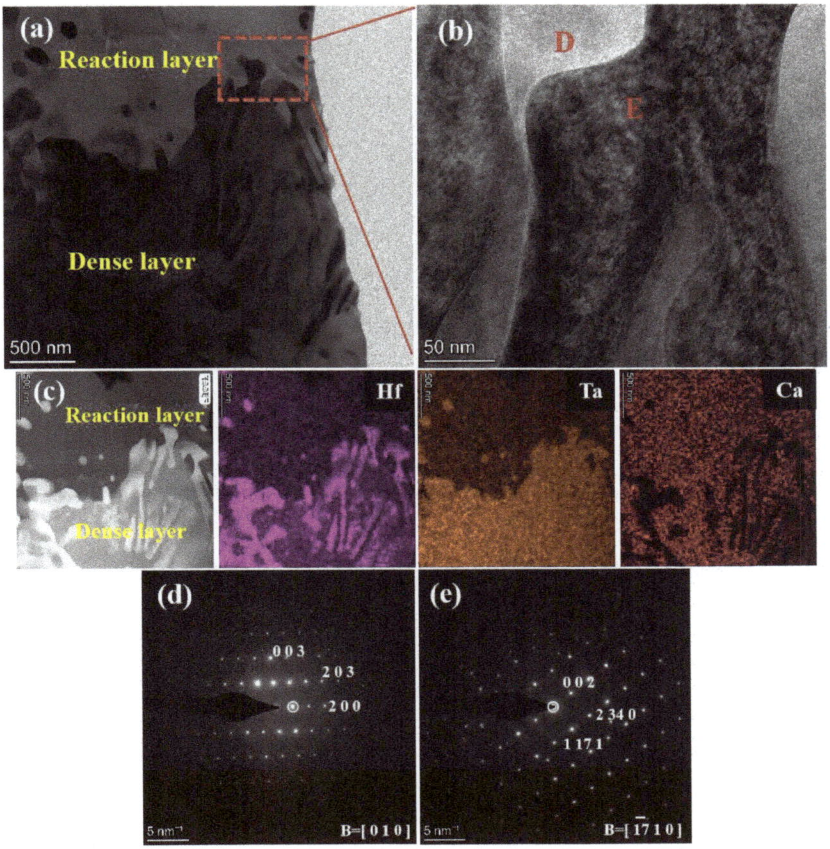

Figure 9. (**a**) Cross-sectional BF TEM image of CMAS reaction front corresponding to the dashed rectangle in Figure 8a; (**b**) magnified HAADF image of reaction interface indicated by in (**a**); (**c**) the corresponding EDS elemental mapping; (**d**,**e**) SAED patterns of D and E in (**b**).

3.3.2. Characterization of Corrosion Products at 1300 °C

Figure 10 shows a cross-sectional SEM image of $Hf_6Ta_2O_{17}$ ceramic corroded by CMAS at 1300 °C for 4 h. A new white bulk product I is formed in the reaction layer, and EDS results in Table 3 confirm that product I is HfO_2 phase. After CMAS attack at 1300 °C for 16 h (Figure 11), the morphology of the reaction layer and the dense layer has changed considerably. The reaction layer mainly consists of blocky phase K and there is a small amount of L phase at the grain boundary. Two kinds of phase structures appear in the dense layer: O and N, and they are cross-embedded with each other, which further enhances the dense layer to prevent the CMAS infiltration. Increasing the corrosion time to 50 and 100 h at 1300 °C, the corrosion products of $Hf_6Ta_2O_{17}$ ceramic show no significant change except the continuous growth of the dense layer. Combining the EDS mapping results in Figure 11d with EDS point analysis in Table 4, K and O are determined to be m-HfO_2 phase. Both the L and N phases are rich in Ca, Hf and Ta elements, but the L phase contains more Hf element and the N phase contains more Ta element. It is speculated that L and N may be the same substance, but the composition is distinct due to different depths. According to Figure 11d, the dense layer prevents Ca element from penetrating into $Hf_6Ta_2O_{17}$ ceramic. Figure 12 presents the XRD patterns of $Hf_6Ta_2O_{17}$ ceramic corroded by CMAS at 1300 °C for 4, 16, 50 and 100 h. The results show that m-HfO_2 has the highest diffraction peak, followed by a new $CaTa_2O_6$ characteristic peak, and a small number of $CaAl_2Si_2O_8$ and

$HfSiO_4$ phases. But there is no significant difference in diffraction peak under different corrosion time. The corrosion products are mainly $CaTa_2O_6$ and m-HfO_2 for different time. Based on the above analysis, the CMAS corrosion products of $Hf_6Ta_2O_{17}$ ceramic at 1300 °C are m-HfO_2 (K, O phase) and $CaTa_2O_6$ (L, N phase).

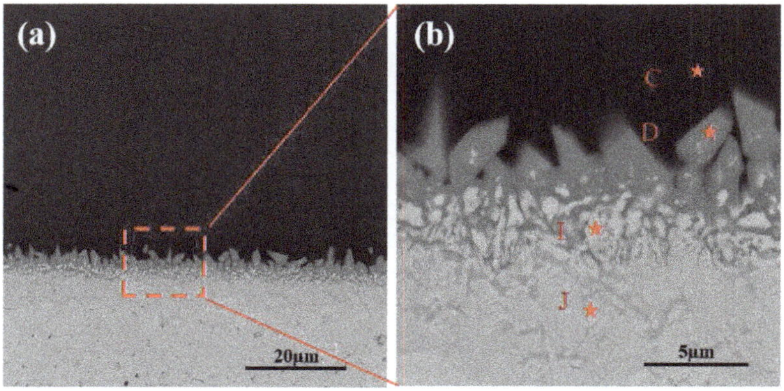

Figure 10. (**a**) A cross-sectional BSE image of $Hf_6Ta_2O_{17}$ ceramic after CMAS corrosion at 1300 °C for 8 h; (**b**) magnified image of the dashed rectangle in (**a**).

Table 3. Chemical compositions of the marked regions in Figure 10b.

Fraction Location	Compositions (at.%)							
	O	Mg	Al	Si	Ca	Hf	Ta	-
C	57.45	0.55	7.99	20.46	12.34	0.69	0.52	CMAS
D	60.76	-	2.17	11.21	12.52	11.55	1.79	$HfSiO_4$
I	65.54	0.01	1.62	3.43	7.51	20.20	1.70	m-HfO_2
J	64.46	-	0.78	-	5.61	23.77	5.38	$Ca_xHf_{6-x}Ta_2O_{17-x}$

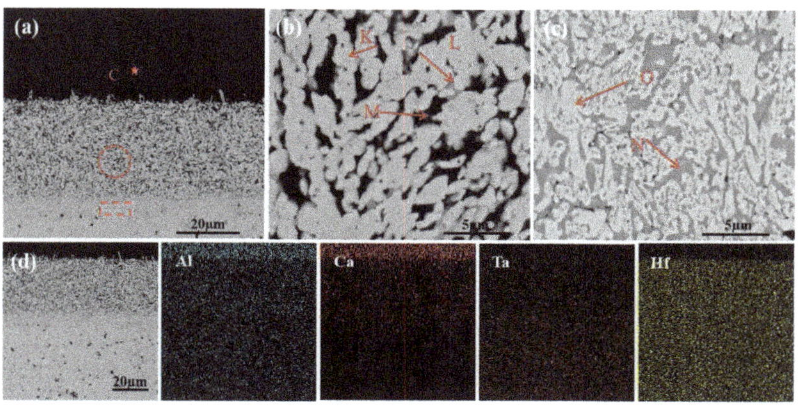

Figure 11. (**a**) A cross-sectional BSE image of $Hf_6Ta_2O_{17}$ ceramic after CMAS corrosion at 1300 °C for 16 h; (**b**) the high magnification BSE image corresponding to the dashed rectangle in (**a**); (**c**) the high magnification BSE image corresponding to the dashed circle in (**a**); (**d**) cross-sectional EDS mapping results of $Hf_6Ta_2O_{17}$ ceramic.

Table 4. Chemical compositions of the marked regions in Figure 11.

Fraction Location	Composition (at.%)						
	O	Mg	Al	Si	Ca	Hf	Ta
C	58.14	3.09	6.13	21.27	8.99	0.91	1.47
K	64.39	0.46	0.90	-	1.58	29.13	3.55
L	63.95	0.08	0.69	2.66	10.42	10.10	12.09
M	59.14	2.36	4.66	19.05	7.87	4.70	2.21
N	56.72	-	0.74	9.32	10.66	3.26	19.29
O	64.86	1.14	0.27	-	0.83	26.98	5.91

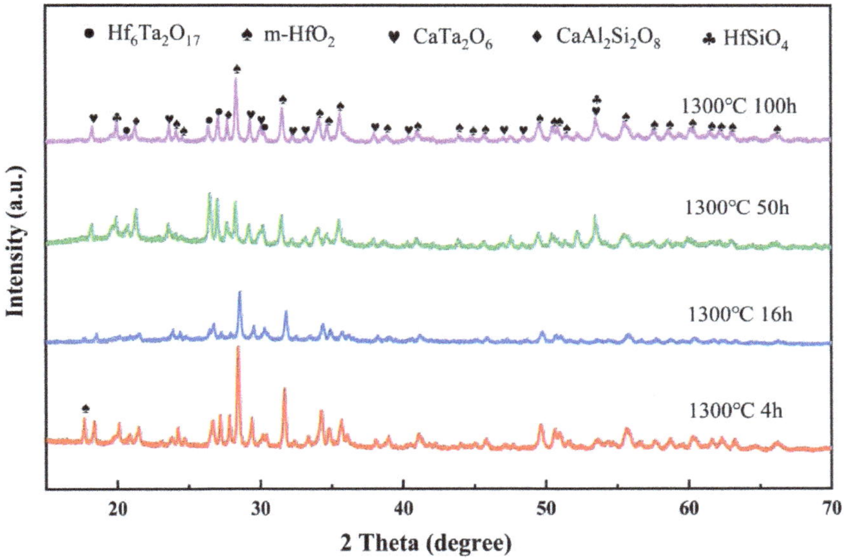

Figure 12. XRD patterns of $Hf_6Ta_2O_{17}$ ceramic after CMAS corrosion at 1300 °C for 4 h, 16 h, 50 h and 100 h.

3.3.3. Characterization of Corrosion Products at 1400 °C

Figure 13 shows cross-sectional SEM images of $Hf_6Ta_2O_{17}$ ceramic corroded by CMAS at 1400 °C for 16 h. The corrosion results are analogous to those of long-time corrosion at 1300 °C, and the cross section is also composed of residual CMAS layer, reaction layer and dense layer. Two notable features are observed at 1400 °C. First of all, there are abundant crystalline phases P in the residual CMAS layer, and new P phase presents a dendritic structure. It is confirmed that P phase is also $CaTa_2O_6$ in Table 5. Secondly, the corrosion depth increases significantly at 1400 °C for 16 h, and the corrosion depth of 140.6 μm is much greater than that at 1300 °C (~34.1 μm). Figure 14 shows XRD patterns of $Hf_6Ta_2O_{17}$ ceramic corroded by CMAS at 1400 °C for 4, 16, 50 and 100 h. The corrosion products are mainly $CaTa_2O_6$ and m-HfO_2. Because of the low XRD detection depth, the $CaTa_2O_6$ phase on the surface of the corroded sample is mainly detected, which is consistent with the cross-sectional BSE results. Based on the XRD and EDS results, the corrosion products are P, R and U phases (m-HfO_2) and T and Y phases ($CaTa_2O_6$) at 1400 °C.

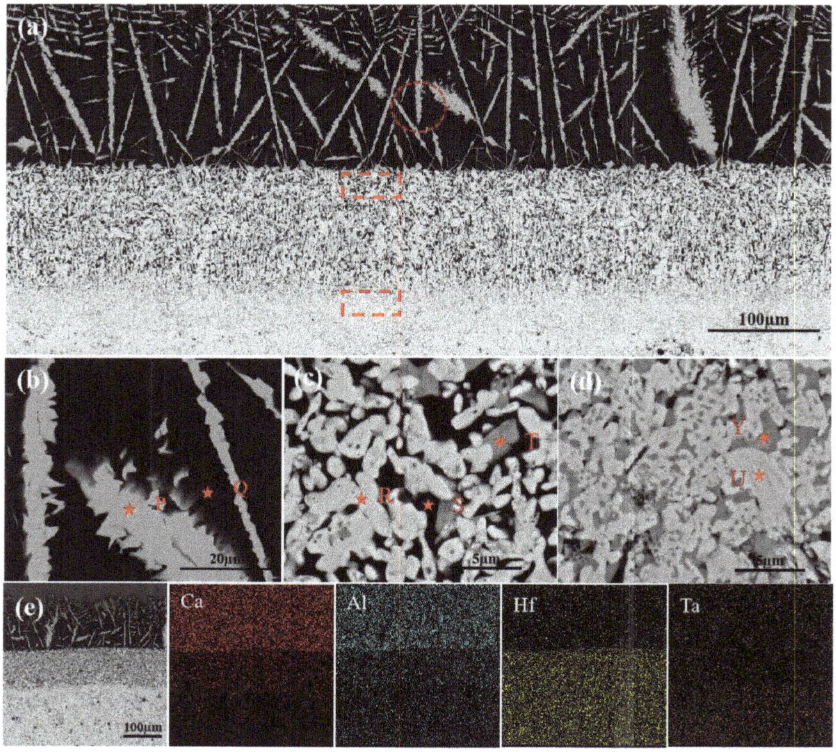

Figure 13. (**a**) A cross-sectional BSE image of $Hf_6Ta_2O_{17}$ ceramic after CMAS corrosion at 1400 °C for 16 h; (**b**) the high magnification BSE image corresponding to the dashed circle in (**a**); (**c**) the high magnification BSE image corresponding to the dashed rectangle in the reaction layer of (**a**); (**d**) the high magnification BSE image corresponding to the dashed rectangle in the dense layer of (**a**); (**e**) cross-sectional EDS mapping results of $Hf_6Ta_2O_{17}$ ceramic.

Table 5. Chemical compositions of the marked regions in Figure 13.

Fraction Location	Composition (at.%)						
	O	Mg	Al	Si	Ca	Hf	Ta
P	58.48	-	1.64	6.96	11.82	1.47	19.63
Q	58.53	0.40	8.07	21.81	9.14	0.71	1.34
R	62.12	0.09	0.58	-	0.44	34.90	1.87
S	59.14	0.21	6.74	20.46	7.99	3.75	1.72
T	58.22	-	1.49	3.45	12.79	10.47	13.59
U	61.95	-	0.53	-	1.37	32.79	3.36
Y	60.72	-	0.25	6.60	10.86	2.69	18.88

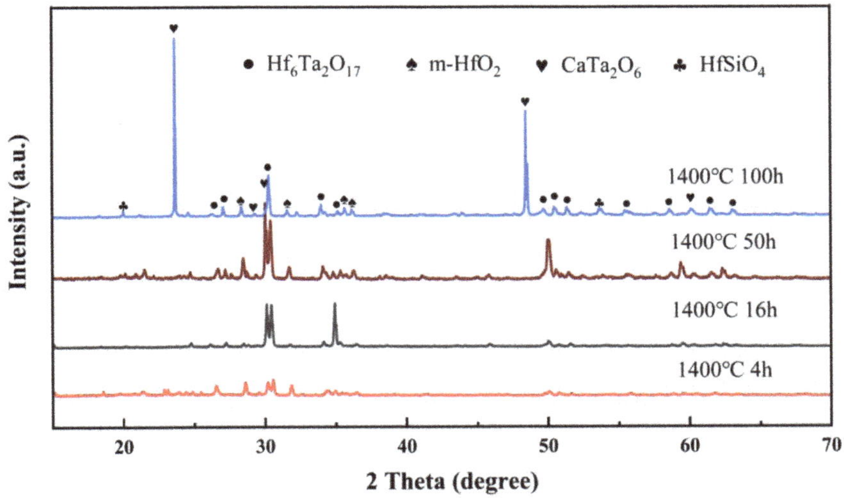

Figure 14. XRD patterns of $Hf_6Ta_2O_{17}$ ceramic after CMAS corrosion at 1400 °C for 4 h, 16 h, 50 h and 100 h.

4. Discussion

4.1. The Excellent CMAS Resistance of $Hf_6Ta_2O_{17}$ Ceramic

Compared to other ceramic material in TBCs, $Hf_6Ta_2O_{17}$ ceramic exhibits excellent CMAS resistance. After 100 h CMAS corrosion at 1250 °C, the CMAS infiltration depth of $Hf_6Ta_2O_{17}$ ceramic only reaches 32.3 μm. In contrast, the CMAS infiltration depths of YSZ and YSZ doped by Er element ceramic at 1250 °C for 2 h reach 21.97 μm and 10.75 μm, respectively [31]. The average CMAS infiltration depths of $NdPO_4$, $SmPO_4$ and $GdPO_4$ ceramic at 1250 °C for 4 h are 140, 60 and 40 μm, respectively [16]. The CMAS infiltration depth of nanosized $Sm_2Zr_2O_7$ ceramic is 47 μm after 48 h CMAS corrosion at 1250 °C [32]. As the corrosion temperature rises to 1300 °C, CMAS resistance of $Hf_6Ta_2O_{17}$ ceramic is superior than other ceramic material in TBCs. The corrosion depths of $Hf_6Ta_2O_{17}$ ceramic after CMAS attack at 1300 °C for 4 and 50 h are 13 and 113.8 μm, respectively. The average thickness of the reaction layer of $La_2(Zr_{0.7}Ce_{0.3})_2O_7$ is 140 μm at 1300 °C for 4 h [17]. For 50 h CMAS exposure at 1300 °C, the CMAS infiltration depths of 17YSZ and $Gd_2Zr_2O_7$ ceramics are 700 and 140 μm, respectively [9,30].

4.2. Effect of Temperature and Time on CMAS Corrosion Resistance of $Hf_6Ta_2O_{17}$ Ceramic

Temperature and time are important factors affecting the corrosion behavior of $Hf_6Ta_2O_{17}$ ceramic. Figure 15 shows the relative change rates of the CMAS infiltration depth with the relative increasing rates of temperature and time. The relative increasing rate x of influencing factors and the change of CMAS infiltration depth y can be expressed by Equations (2) and (3), respectively [33]:

$$x = \frac{x_i - x_0}{x_0} \times 100\% \tag{2}$$

$$y = \frac{y(x_i) - y(x_0)}{y(x_0)} \times 100\%, \tag{3}$$

where x_i and x_0 represent the immediate value and the initial value of an influencing factor, respectively, and $y(x_0)$ and $y(x_i)$ are the infiltration depth of CMAS at x_i and x_0, respectively. It can be seen that temperature is the most important factor affecting CMAS infiltration, while time has a less effect on CMAS infiltration depth. A 4% increase in temperature leads to a 278% increase in the CMAS infiltration depth. When the time increases by 100%, the infiltration depth only increases by only 38%.

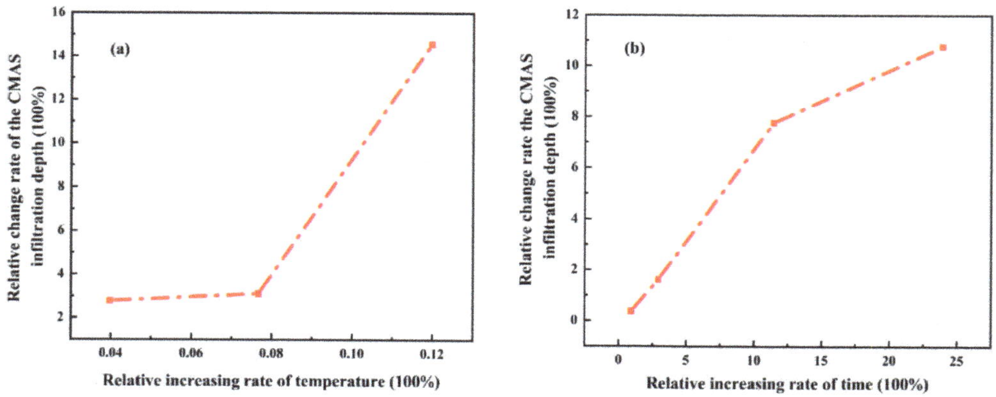

Figure 15. Relative change rates of the CMAS infiltration depth with the relative increasing rates of temperature (**a**) and time (**b**).

Considerable studies have proven that CMAS viscosity decreases with an increase in temperature, and the CMAS infiltration rate increases with an increase in temperature. CMAS viscosity is directly related to the CMAS infiltration depth in TBCs at high temperature [8,34,35]. CMAS viscosity at different temperatures can be calculated by using the Vogel–Fulcher–Tammann (VFT) equation [36]:

$$\log\eta = A + \frac{B}{T - C} \tag{4}$$

where η is viscosity, A is a constant, B and C are related to the composition of CMAS, and T is the temperature. The parameter A is assumed to be a constant, which is the value of $\log\eta$ at infinite temperature. B and C are controlled by the CMAS composition effect, which can be estimated by using Equations (5) and (6) [8,37]:

$$B = \sum_{i=1}^{7}[b_i M_i] + \sum_{j=1}^{3}[b_{1j}(M1_{1j} \cdot M2_{1j})] \tag{5}$$

$$C = \sum_{i=1}^{6}[c_i N_i] + [c_{11}(N1_{11} \cdot N2_{11})], \tag{6}$$

where M_s and N_s refer to the combination of mol% oxides and the unknown coefficients b_i, b_{1j}, c_i, c_{11} are adequate to compute the values of B and C for any individual melt composition.

According to Equation (4), an increase in corrosion temperature can reduce the CMAS viscosity by an order of magnitude, leading to an increase in the CMAS infiltration depth. Consequently, temperature is the most important factor in determining the CMAS corrosion behavior.

4.3. CMAS Resistance Mechanism of $Hf_6Ta_2O_{17}$ Ceramic

Since temperature is the main factor affecting the corrosion behavior of $Hf_6Ta_2O_{17}$ ceramic, the CMAS resistance mechanism is divided into two parts: the low-temperature section at 1250 °C and the high-temperature section at 1300–1400 °C. At low temperature, the reaction between $Hf_6Ta_2O_{17}$ ceramic and CMAS is not high. The reaction layer has only a small amount of $HfSiO_4$, and mainly residual CMAS is self-crystallized to form wollastonite and anorthite, as shown in reaction Equations (7) and (8). This is because a large amount of Ca element infiltrates into the dense layer, while Al element is blocked in the residual CMAS layer, resulting in changes in the composition of residual CMAS and the easier formation of self-crystallizing products [14]. $Hf_6Ta_2O_{17}$ ceramic has excellent CMAS corrosion resistance at 1250 °C, which can be attributed to two aspects. Firstly, $Hf_6Ta_2O_{17}$ ceramic at 1250 °C has excellent chemical inertness under high temperature, so the molten CMAS is less destructive to the ceramic. The formation of CMAS self-crystallization products consumes much of the Ca and Al elements, increasing the CMAS viscosity, thus impeding the downward infiltration of CMAS. Secondly, the appearance of a reaction layer and a dense layer effectively inhibits the downward infiltration of molten CMAS; in particular, the Ca element penetrates the $HfSiO_4$ reaction layer and continues to penetrate downward to form the $Ca_xHf_{6-x}Ta_2O_{17-x}$ dense layer, which suppresses CMAS above the ceramic. SEM and TEM results also indicate that the Ca elements penetrate downward through volume diffusion rather than grain boundary corrosion, thus greatly delaying the CMAS infiltration.

$$2CaO(CMAS) + Al_2O_3(CMAS) + 3SiO_2(CMAS) \rightarrow CaAl_2Si_2O_8 + CaSiO_3 \quad (7)$$

$$Hf_6Ta_2O_{17} + xCaO(CMAS) + xSiO_2(CMAS) \rightarrow xHfSiO_4 + Ca_xHf_{6-x}Ta_2O_{17-x} \quad (8)$$

The corrosion behavior of $Hf_6Ta_2O_{17}$ ceramic is greatly changed at high temperatures, and the most remarkable feature is the formation of m-HfO_2 and $CaTa_2O_6$. The CMAS corrosion behavior of $Hf_6Ta_2O_{17}$ ceramic needs to be divided into two stages at 1300 °C: (I) $t < 16$ h and (II) 16 h $\leq t \leq$ 100 h. During stage I, a small amount of m-HfO_2 is precipitated from the reaction layer, and the corrosion results of the ceramic are basically consistent with that at 1250 °C. During stage II, both the reaction layer and the dense layer are composed of $CaTa_2O_6$ and m-HfO_2. The thickness of the reaction layer is basically unchanged, while the thickness of the dense layer keeps increasing at long-term corrosion. Recent research has shown that m-HfO_2 and $CaTa_2O_6$ have extremely low solubility in molten CMAS and will not generate reaction products, which also suggests that m-HfO_2 and $CaTa_2O_6$ can form dense layers that hinder downward infiltration of CMAS [38,39]. The corrosion behavior of $Hf_6Ta_2O_{17}$ ceramic at 1400 °C is similar to that of stage II at 1300 °C. The difference is that a large amount of Ta element begins to diffuse upward to the residual CMAS layer, resulting in the formation of $CaTa_2O_6$. Therefore, the corrosion behavior of $Hf_6Ta_2O_{17}$ ceramic at 1300–1400 °C can be explained from two aspects. Firstly, in terms of thermodynamics, $Hf_6Ta_2O_{17}$ ceramic reacts violently with molten CMAS to generate m-HfO_2 and $CaTa_2O_6$ at 1300–1400 °C, as shown in reaction Equation (9). The Ca element of molten CMAS plays an important role in the reaction process at this stage. The corrosion kinetics can also explain the corrosion resistance mechanism of $Hf_6Ta_2O_{17}$ ceramic. According to Figures 4 and 5, CMAS corrosion for a long time is not determined by chemical reaction, but penetration downward through solid diffusion. Continuous thickening of the dense layer can effectively improve the corrosion resistance of $Hf_6Ta_2O_{17}$ ceramic. At the initial stage of CMAS corrosion at 1400 °C, the CMAS infiltration depth increases sharply, because the time is too short to generate a thick dense layer. Over time, the dense layer continues to thicken, and the corrosion rate begins to decrease significantly.

$$Hf_6Ta_2O_{17} + CaO(CMAS) \rightarrow 6HfO_2 + CaTa_2O_6 \quad (9)$$

5. Conclusions

In this paper, the CMAS corrosion behavior of $Hf_6Ta_2O_{17}$ ceramic at 1250–1400 °C is investigated, the effects of temperature and time on the CMAS corrosion resistance of $Hf_6Ta_2O_{17}$ ceramic are discussed, and the CMAS resistance mechanism is clarified. Several conclusions can be drawn, as follows:

(1) Compared with some traditional and novel CMAS-resistant ceramic materials in TBCs, $Hf_6Ta_2O_{17}$ ceramic exhibits excellent CMAS resistance characteristics at short-term and long-term corrosion;

(2) Based on the influence of CMAS viscosity, temperature is the most important factor affecting the CMAS behavior of $Hf_6Ta_2O_{17}$ ceramic. At 1250 °C, CMAS self-crystallization products are composed of anorthite $CaAl_2Si_2O_8$ and wollastonite $CaSiO_3$. The reaction layer is composed of $HfSiO_4$ and the dense layer is consisted of $Ca_xHf_{6-x}Ta_2O_{17-x}$. At 1300 and 1400 °C, the reaction layer and the dense layer are composed of $CaTa_2O_6$ and m-HfO_2.

(3) The CMAS resistance mechanism of $Hf_6Ta_2O_{17}$ ceramic varies with temperature. At 1250 °C, the formation of CMAS self-crystallization products consisting of anorthite $CaAl_2Si_2O_8$ and wollastonite $CaSiO_3$, and the formation of $HfSiO_4$ in a reaction layer and $Ca_xHf_{6-x}Ta_2O_{17-x}$ in a dense layer effectively inhibit the CMAS infiltration. At 1300 and 1400 °C, the formation and thickening of $CaTa_2O_6$ and m-HfO_2 in the dense layer improve the CMAS corrosion resistance of $Hf_6Ta_2O_{17}$ ceramic.

Author Contributions: Conceptualization, W.Z.; formal analysis, X.H. and F.Z.; funding acquisition, W.Z. and J.X.; methodology, S.L., Q.L., X.H., W.Z. and F.Z.; resources, J.G. and J.X.; supervision, J.G. and W.Z.; validation, X.H.; writing—original draft, S.L. and Q.L.; writing—review & editing, W.Z. and F.Z. All authors have read and agreed to the published version of the manuscript.

Funding: This work was supported by the Science and Technology Innovation Program of Hunan Province (Grant No. 2022RC1082), the Scientific Research Foundation of Hunan Provincial Education Department (Grant No. 21A0120), the National Natural Science Foundation of China (Grant No. 52171015), the Natural Science Foundation of Hunan Province (Grant No. 2021JJ30467), and Postgraduate Scientific Research Innovation Project of Hunan Province (Grant No. XDCX2021B139).

Institutional Review Board Statement: Not applicable.

Informed Consent Statement: Not applicable.

Data Availability Statement: The raw/processed data required to reproduce these findings cannot be shared at this time as the data also forms part of an ongoing study.

Conflicts of Interest: The authors declare no conflict of interest.

References

1. Clarke, D.; Oechsner, M.; Padture, N. Thermal-barrier coatings for more efficient gas-turbine engines. *MRS Bull.* **2012**, *37*, 891–899. [CrossRef]
2. Vasile, B.; Birca, A.; Surdu, V.; Neacsu, I.; Nicoară, A. Ceramic composite materials obtained by electron-beam physical vapor deposition used as thermal barriers in the aerospace industry. *Nanomaterials* **2020**, *10*, 370. [CrossRef] [PubMed]
3. Darolia, R. Thermal barrier coatings technology: Critical review, progress update, remaining challenges and prospects. *Int. Mater. Rev.* **2013**, *58*, 315–348. [CrossRef]
4. Rabiei, A.; Evans, A. Failure mechanisms associated with the thermally grown oxide in plasma-sprayed thermal barrier coatings. *Acta Mater.* **2000**, *48*, 3963–3976. [CrossRef]
5. Chen, X.; Wang, R.; Yao, N.; Evans, A.; Hutchinson, J.; Bruce, R.W. Foreign object damage in a thermal barrier system: Mechanisms and simulations. *Mater. Sci. Eng. A* **2003**, *352*, 221–231. [CrossRef]
6. Levi, C.; Hutchinson, J.; Vidal-Sétif, M.; Johnson, C. Environmental degradation of thermal-barrier coatings by molten deposits. *MRS Bull.* **2012**, *37*, 932–941. [CrossRef]
7. Krause, A.; Garces, H.; Dwivedi, G.; Ortiz, A.; Sampath, S. Calcia-magnesia-alumino-silicate (CMAS)-induced degradation and failure of air plasma sprayed yttria-stabilized zirconia thermal barrier coatings. *Acta Mater.* **2016**, *105*, 355–366. [CrossRef]
8. Nieto, A.; Agrawal, R.; Bravo, L.; Hofmeister-Mock, C.; Pepi, M.; Ghoshal, A. Calcia-magnesia-alumina-silicate (CMAS) attack mechanisms and roadmap towards Sandphobic thermal and environmental barrier coatings. *Int. Mater. Rev.* **2021**, *66*, 451–492. [CrossRef]
9. Pujol, G.; Ansart, F.; Bonino, J.; Malié, A.; Hamadi, S. Step-by-step investigation of degradation mechanisms induced by CMAS attack on YSZ materials for TBC applications. *Surf. Coat. Technol.* **2013**, *237*, 71–78. [CrossRef]
10. Morelli, S.; Testa, V.; Bolelli, G.; Ligabue, O.; Molinari, E.; Antolotti, N.; Lusvarghi, L. CMAS corrosion of YSZ thermal barrier coatings obtained by different thermal spray processes. *J. Eur. Ceram. Soc.* **2020**, *40*, 4084–4100. [CrossRef]
11. Yan, Z.; Guo, L.; Zhang, Z.; Wang, X.; Ye, F. Versatility of potential protective layer material Ti_2AlC on resisting CMAS corrosion to thermal barrier coatings. *Corros. Sci.* **2020**, *167*, 108532. [CrossRef]
12. Yang, W.; Ye, F. The thermophysical properties and the molten CMAS resistance performance of Ytterbium Tantalate. *Surf. Coat. Technol.* **2021**, *423*, 127584. [CrossRef]

13. Gok, M.; Goller, G. Microstructural characterization of GZ/CYSZ thermal barrier coatings after thermal shock and CMAS+ hot corrosion test. *J. Eur. Ceram. Soc.* **2017**, *37*, 2501–2508. [CrossRef]
14. Qu, W.; Li, S.; Chen, Z.; Li, C.; Pei, Y.; Gong, S. Hot corrosion behavior and wettability of calcium-magnesium-alumina-silicate (CMAS) on $LaTi_2Al_9O_{19}$ ceramic. *Corros. Sci.* **2020**, *162*, 108199. [CrossRef]
15. Krause, A.; Garces, H.; Herrmann, C.; Padture, N. Resistance of $2ZrO_2{\cdot}Y_2O_3$ top coat in thermal/environmental barrier coatings to calcia-magnesia-aluminosilicate attack at 1500 °C. *J. Am. Ceram. Soc.* **2017**, *100*, 3175–3187. [CrossRef]
16. Wang, F.; Guo, L.; Wang, C.; Ye, F. Calcium-magnesium-alumina-silicate (CMAS) resistance characteristics of $LnPO_4$ (Ln= Nd, Sm, Gd) thermal barrier oxides. *J. Eur. Ceram. Soc.* **2017**, *37*, 289–296. [CrossRef]
17. Zhou, X.; Zou, B.; He, L.; Xu, Z.; Xu, J.; Mu, R.; Cao, X. Hot corrosion behaviour of $La_2(Zr_{0.7}Ce_{0.3})_2O_7$ thermal barrier coating ceramics exposed to molten calcium magnesium aluminosilicate at different temperatures. *Corros. Sci.* **2015**, *100*, 566–578. [CrossRef]
18. Chen, L.; Hu, M.; Guo, J.; Chong, X.; Feng, J. Mechanical and thermal properties of $RETaO_4$ (RE= Yb, Lu, Sc) ceramics with monoclinic-prime phase. *J. Mater. Sci. Technol.* **2020**, *52*, 20–28. [CrossRef]
19. Clarke, D.; Phillpot, S. Thermal barrier coating materials. *Mater. Today* **2005**, *8*, 22–29. [CrossRef]
20. Barsoum, M.; Salama, I.; El-Raghy, T.; Golczewski, J.; Seifert, H.; Aldinger, F.; Porter, W.; Wang, H. Thermal and electrical properties of Nb_2AlC, (Ti, $Nb)_2AlC$ and Ti_2AlC. *Metall. Mater. Trans. A* **2002**, *33*, 2775–2779. [CrossRef]
21. Mccormack, S.; Kriven, W. Crystal structure solution for the $A_6B_2O_{17}$ (A = Zr, Hf; B = Nb, Ta) superstructure. *Acta Crystallogr. Sect. B Struct. Sci.* **2019**, *75*, 227–234. [CrossRef]
22. Liu, Q.; Hu, X.; Zhu, W.; Guo, J.; Tan, Z. Effects of Ta_2O_5 content on mechanical properties and high-temperature performance of $Zr_6Ta_2O_{17}$ thermal barrier coatings. *J. Am. Ceram. Soc.* **2021**, *104*, 6533–6544. [CrossRef]
23. McCormack, S.; Tseng, K.; Weber, R.; Kapush, D.; Ushakov, S.; Navrotsky, A.; Kriven, W. In-situ determination of the HfO_2-Ta_2O_5 temperature phase diagram up to 3000 °C. *J. Am. Ceram. Soc.* **2019**, *102*, 4848–4861. [CrossRef]
24. McCormack, S.; Weber, R.; Kriven, W. In-situ investigation of $Hf_6Ta_2O_{17}$ anisotropic thermal expansion and topotactic, peritectic transformation. *Acta Mater.* **2018**, *161*, 127–137. [CrossRef]
25. Li, M.; Xu, Q.; Zhu, S.; Wang, L. Preparation and thermal conductivity of $Hf_6Ta_2O_{17}$ ceramic. *Key Eng. Mater.* **2010**, *434–435*, 459–461.
26. Tan, Z.; Yang, Z.; Zhu, W.; Yang, L.; Zhou, Y.; Hu, X. Mechanical properties and calcium-magnesium-alumino-silicate (CMAS) corrosion behavior of a promising $Hf_6Ta_2O_{17}$ ceramic for thermal barrier coatings. *Ceram. Int.* **2020**, *46*, 25242–25248. [CrossRef]
27. Zhao, H.; Levi, C.; Wadley, H. Molten silicate interactions with thermal barrier coatings. *Surf. Coat. Technol.* **2014**, *251*, 74–86. [CrossRef]
28. Borom, M.P.; Johnson, C.A.; Peluso, L.A. Role of environmental deposits and operating surface temperature in spallation of air plasma sprayed TBC. *Surf. Coat. Technol.* **1996**, *86*, 116–126. [CrossRef]
29. Zaleski, E.; Ensslen, C.; Levi, C. Melting and crystallization of silicate systems relevant to thermal barrier coating damage. *J. Am. Ceram. Soc.* **2015**, *98*, 1642–1649. [CrossRef]
30. Wu, D.; Yao, Y.; Shan, X.; Cai, H.; Hu, L.; Zhao, X. ZrO_2-doped $YTaO_4$ as potential CMAS-resistant materials for thermal barrier coatings application. *J. Am. Ceram. Soc.* **2021**, *104*, 6029–6043. [CrossRef]
31. Mo, Y.; Zhang, Y.; Chen, Z.; Zheng, H.; Li, G.; Gen, Y.; Li, G.; Peng, P. Effect of Er dopant on the corrosion resistance of YSZ in CMAS melt: Experimental and first-principles study. *J. Mater. Sci.* **2021**, *56*, 17542–17555. [CrossRef]
32. Wang, Y.; Ma, Z.; Liu, L.; Liu, Y. Influence of the grain size on CMAS attack of $Sm_2Zr_2O_7$ ceramic. *Ceram. Int.* **2021**, *47*, 24453–24457. [CrossRef]
33. Cao, L.; Hu, H.; Gao, D. Design and fabrication of micro-textures for inducing a superhydrophobic behavior on hydrophilic materials. *Langmuir* **2007**, *23*, 4310–4314. [CrossRef]
34. Poerschke, D.; Jackson, R.; Levi, C. Silicate deposit degradation of engineered coatings in gas turbines: Progress toward models and materials solutions. *Annu. Rev. Mater. Res.* **2017**, *47*, 297–330. [CrossRef]
35. Kumar, R.; Rommel, S.; Jiang, C.; Jordan, E. Effect of CMAS viscosity on the infiltration depth in thermal barrier coatings of different microstructures. *Surf. Coat. Technol.* **2022**, *432*, 128039. [CrossRef]
36. Fulcher, G.S. Analysis of recent measurements of the viscosity of glasses. *J. Am. Ceram. Soc.* **1925**, *8*, 339–355. [CrossRef]
37. Giordano, D.; Russell, J.; Dingwell, D. Viscosity of magmatic liquids: A model. *Earth Planet. Sci. Lett.* **2008**, *271*, 123–134. [CrossRef]
38. Holgate, C.; Yang, Y.; Levi, C. Reactive crystallization in HfO_2 exposed to molten silicates. *J. Eur. Ceram. Soc.* **2021**, *41*, 5686–5695. [CrossRef]
39. Li, B.; Chen, L.; Luo, K.; Guo, J.; Wang, J.; Zhang, L.; Feng, J. Synthesis and thermophysical properties of ATa_2O_6 (A= Co, Ni, Mg, Ca) tantalates with robust CMAS resistance. *J. Am. Ceram. Soc.* **2022**, *105*, 4840–4858. [CrossRef]

Disclaimer/Publisher's Note: The statements, opinions and data contained in all publications are solely those of the individual author(s) and contributor(s) and not of MDPI and/or the editor(s). MDPI and/or the editor(s) disclaim responsibility for any injury to people or property resulting from any ideas, methods, instructions or products referred to in the content.

coatings

Article

The Influence of H Content on the Properties of a-C(W):H Coatings

Manuel Evaristo [1,*], Filipe Fernandes [1,2], Chris Jeynes [3] and Albano Cavaleiro [1,4]

1 CEMMPRE—Centre for Mechanical Engineering Materials and Processes, Department of Mechanical Engineering, University of Coimbra, Rua Luís Reis Santos, 3030-788 Coimbra, Portugal
2 ISEP—School of Engineering, Polytechnic of Porto, Rua Dr. António Bernardino de Almeida 431, 4200-072 Porto, Portugal
3 Ion Beam Centre, Advanced Technology Institute, University of Surrey, Guildford GU2 7XH, UK
4 IPN-LED&MAT—Instituto Pedro Nunes, Laboratório de Ensaios, Desgaste e Materiais, Rua Pedro Nunes, 3030-199 Coimbra, Portugal
* Correspondence: manuel.evaristo@dem.uc.pt; Tel.: +351-239793745

Abstract: Diamond-like-carbon "DLC" coatings can be deposited in many different ways, giving a large range of material properties suitable for many different types of applications. Hydrogen content significantly influences the mechanical properties and the tribological behavior of DLC coatings, but its determination requires techniques that are not available in many research centers. Thus, it is important to find alternative indirect techniques, such as Raman spectroscopy or nanoindentation (hardness measurements), which can give comparative and indicative values of the H contents in the coatings, particularly when depositions with a reactive gas flow are being studied. In this work, "DLC" composite coatings with varying H content were deposited via Physical Vapor Deposition (PVD) magnetron sputtering in a reactive atmosphere (Ar + CH_4). An Ion Beam Analysis was used to determine the elemental depth profile across the coating thickness (giving both average C:W:H ratios and film density when combined with profilometer measurements of film thickness). The hardness was evaluated with nanoindentation, and a decrease from 16 to 6 GPa (and a decrease in the film density by a factor of two) with an increasing CH_4 flow was observed. Then, the hardness and Raman results were correlated with the H content in the coatings, showing that these indirect methods can be used to find if there are variations in the H content with the increase in the CH_4 flow. Finally, the adhesion and tribological performance of the coatings were evaluated. No significant differences were found in the adhesion as a function of the H content. The tribological properties presented a slight improving trend with the increase in the H content with a decrease in the wear rate and friction.

Keywords: DLC coatings; metal doped; hydrogen content; Raman spectroscopy; mechanical properties; ERD; film density

1. Introduction

Diamond-like carbon (DLC) coatings are used today in a wide range of applications where low friction and wear are required [1]. Composite DLC coatings can have greatly improved tribological properties, and tungsten-doped DLC coatings have been well reviewed by Andrzej Czyzniewski [2,3]. DLC is an amorphous carbon (a-C) where the bonding is pure sp^3 (valence 4, as diamond). Hydrogen is added (a-C:H) to relieve strain, and usually this results in some sp^2 (polymer) bonding, which also modifies the hardness. Strictly speaking, this modified material is no longer entirely "diamond-like", and the more it is modified the less diamond-like it is, but it is still normally referred to as "DLC". Here, we study the effects of increasing the H-content of W-doped DLC ("a-C(W):H").

In addition to the different manufacturing techniques and various doping elements (which can have a major influence on the behavior of DLC coatings [4–7]), hydrogen has a great impact on the film's character, stabilizing the covalent bonding network (sp^3) and controlling the optical and electrical properties. Moreover, hydrogen plays a key role in the mechanical and tribological behavior of the coatings. The influence on the coating's hardness is significant, but this is a complex phenomenon depending on many factors including the precursor gases used, the selected deposition process and the processing conditions [8]. Some authors found a direct correlation between the H content and the mechanical properties, hardness and Young's modulus, showing a decrease in the hardness as the H content in the coatings increases. Casiraghi et al. [9] found an increase in the Young's modulus with the increase in the value of one Raman parameter related with the type of bonds and the clustering in the coating.

However, there are other factors that must also be considered. The deposition method influences the mechanical properties of the coatings; coatings with identical H content can have different hardness values [10,11]. Additionally, the precursor gas used influences the coating's main characteristics, and coatings deposited with different precursor gases can have different mechanical properties [12,13]. Thus, the use of different deposition methods or precursor gases can significantly influence the type of bonding in the coatings as well as the structure and morphology; as a consequence, different mechanical properties can be achieved [14]. In general, if all the deposition conditions are kept constant, with the exception of the one determining the H content, one expects a decrease in the hardness as the H content increases.

Raman spectroscopy is one of the easiest ways to obtain structural information for different forms of carbon materials, from crystalline to amorphous [15]. In most cases of DLC coatings deposited by sputtering, the spectra of carbon-based coatings show two prominent features, the D band around 1300–1400 cm^{-1} and the G band around 1500–1600 cm^{-1}, and with UV radiation, a T peak can also be observed around 980–1060 cm^{-1} [16]. The D band is due to the "breathing modes" of sp^2 atoms in distorted rings and the G band is related to the stretching of pairs of sp^2 atoms in rings and chains [17]. Raman provides information (i) about the coating's structure through the G peak position, (ii) the intensity ratio between the D and G bands (ID/IG ratio) and (iii) the full width at a half maximum of the G band (FWHM (G)).

The photoluminescence background in visible Raman spectroscopy is associated with the H presence in DLC coatings and can be used to quantify the H content [9]. However, this is not straightforward, since many other factors also influence the photoluminescence background, such as the metal content or the presence of oxygen in the coatings [18].

In this study a set of a-C(W):H coatings was deposited with constant conditions, only varying the CH_4 flow between 10 and 40 sccm. First, the H content of the films is correlated with Raman and hardness. Then, the adhesion and tribological properties of the coatings are studied. The elemental depth profiles (including H) were obtained absolutely using a detailed Ion Beam Analysis [19].

2. Materials and Methods

The coatings were deposited with a four magnetrons UDP-650-4 Teer Coatings, Worcestershire, UK equipment with DC power supplied to the targets and pulsed-DC for biasing the substrates. Four targets were used: one pure chromium, one graphite with 14 pellets of W (20 mm diameter) placed in the preferential erosion zone and two pure graphite targets. The targets were supplied by Testbourne Ltd., Basingstoke, UK with a purity of 99.5%. Before the depositions, the targets and the substrates were sputtered cleaned during 40 min with shutters interposed between them to avoid cross-contamination. To improve the adhesion of the coatings, a Cr-based interlayer of approximately 300 nm was deposited. In the first 10 min, only the Cr target sputtered, followed by 10 min of transition for the deposition of the coating. In this step, the power on the C targets gradually increased as the Cr target power decreased. The coatings were deposited with a rotation speed of the substrates high enough to avoid the formation of a multilayer structure. A power of 1750 W was applied to both carbon targets whereas the target with W pellets was sputtered with 400 W, and a −50 V bias voltage was applied to the substrates. The gas was introduced with mass flow controllers Aera ROD-4. The Ar flow was set to 46 sccm, giving rise to 3.3×10^{-3} mbar while CH_4 was varied between 10 and 40 sccm, corresponding to partial pressures of 1×10^{-4} and 9×10^{-4} mbar, respectively. The total time, including the deposition of the Cr interlayer, was set to 150 min. In all depositions, the chamber was evacuated down to a base pressure better than $\leq 6 \times 10^{-6}$ mbar.

The coatings were deposited on heat-treated AISI M2 steel with a hardness of ~9 GPa, a diameter of 50 mm and a thickness of 5 mm, to achieve the mirror finishing of the surfaces to be coated, and the steel samples were polished with SiC papers (from P120 grit size) followed by an adequate diamond suspension of particles that were 3 μm. Silicon wafers that were polished one side were also used as substrates for the hardness and IBA measurements. The Si and steel substrates were deposited simultaneously. The linear film thicknesses were obtained with a mechanical profilometer of a step in a silicon-coated sample by removing a small drop of boron nitride to have a step corresponding to the total coating thickness. The equipment used to measure the step was the Perthometer S4P, Mahr Perthen GmbH, Hannover, Germany.

A Raman microspectrometer (Xplora, HORIBA Jobin-Yvon, Edison, NJ, USA) with a laser wavelength of 532 nm was used to characterize the structure of the coatings. The laser power was kept below 0.25 mW to avoid any damage to the coatings surface due to heating, which can lead to the graphitization of the coating. For each sample, two spectra were acquired. The Raman spectra were fitted with symmetric Gaussian functions with a linear background in the range 800 to 2000 cm^{-1}.

The nanoindentation measurements were performed on a Nano Test Platform 1 (Micro Materials, Wrexham, UK) with a Berkovich indenter with a load of 5 mN and 30 s to increase and decrease the load and to also maintain the maximum load. The 5 mN maximum load was selected to prevent a maximum penetration depth higher than 10% of the total coating thickness in order to avoid the influence of the Si substrate on the measurements [20]. The Young's modulus (E) of the coatings was determined from the reduced value assumed for the diamond indenter E_i = 1140 GPa and a Poisson ratio, ν_I = 0.07, and ν_s = 0.2 for the coating.

The adhesion of the coatings was evaluated with a conventional scratch tester (CSEM Revetest, Neuchâtel, Switzerland) with a diamond tip (Rockwell, Milwaukee, WI, 200 μm radius) with a loading rate of 10 N/mm to a maximum load of 50 N. For each sample, at least two tests were performed, and the scratch tracks were observed with an optical microscope to determine the load for the first adhesive failure (LC2).

The elemental depth profiles (composition) of the coatings in units of areal mass density (mass/area) were obtained explicitly using Ion Beam Analysis (IBA) at the Surrey Ion Beam Centre: the Cr signal (and the Si substrate signal) terminated the profile, which could therefore be integrated to obtain the total film thickness (as mass/area). Therefore, if the linear film thickness is known (in μm), the film density (mass/volume) can be obtained. Further details about the procedures used are described in the Supplementary Materials.

Two ion beams (3045 keV $^4He^+$ and 4315 keV $^4He^{++}$), four detectors (three particle detectors at scattering angles of 172.8°, 148.6° and 30.75° where the forward recoil detector had a 32 μm Kapton range foil and an X-ray detector with a 146 μm Be filter at 60° to the beam) and two beam incidence angles (normal incidence and glancing incidence) were used. The recoiled H was only observed in the forward recoil detector at the glancing beam incidence.

The IBA therefore included elastic scattering (both Rutherford backscattering, RBS, and non-Rutherford elastic backscattering, EBS, where the EBS scattering cross-sections were obtained from SigmaCalc [21]) elastic recoil detection analysis (ERD) and PIXE. All these data were fitted self-consistently using the DataFurnace code [22] (with NDFv9.4e [23] as the computation engine). The particle-induced X-ray emission (PIXE) spectra had significant signals for W, Ar, Cr and Si, and the thickness of the Cr interlayer was determined from the PIXE, where the signal from the graded interfaces was also fitted by the code. The H content of the films was directly observed with ERD in the forward recoil particle detector. The EBS signals for O and C were greatly enhanced at the resonances at 3038 keV (for O) and 4315 keV (for C). Therefore, this IBA had good sensitivity for all the signals reported.

The tribological behavior was evaluated using a pin-on-disc tribometer (High Temperature tribometer, CSM Instruments, Peseux, Switzerland) in room conditions. The sliding partner was a 6 mm AISI 52110 steel ball. The selection of a steel counterpart is related to the types of applications to which this type of coating is applied. The linear speed, normal load and number of cycles were identical for all samples at 0.1 m s^{-1}, 10 N and 10.000, respectively. The friction coefficients are the mean values of the entire friction curve, and the 'running-in' phase was not considered. The wear rates were calculated as the worn volume per sliding distance per normal load. The worn volumes were measured with an optical 3D profilometer white light interferometer (NewView 7200, Zygo, Middlefield, CT, USA).

3. Results and Discussion

Table 1 shows the average elemental composition of the films integrated from the IBA depth profiles, including samples with six different CH_4 flow rates (one sample with excessive O content was excluded). Additionally shown is the deposition rate (in nm/s) and the film density (in g/cm^3) obtained by combining the IBA thickness (in μg/cm^2) and the thickness (in μm) measured with the profilometer. The a-C(W):H film signals were easily separated from the substrate by the Cr interlayer, whose thickness was determined by PIXE. These IBA spectra were complicated, and for a good fit, it was also necessary to fit the graded Cr interlayer with an equivalent Cr thickness (around 0.3 μm for all the samples). Note that reliable results require good fitting of all the signals from all four detectors. A full uncertainty analysis was not carried out, but the uncertainty was conservatively estimated at about 10% (±5%).

Table 1. Coating composition (by IBA) and density (by IBA and profilometer) together with deposition rate, all as a function of methane flow rate.

CH_4 (sccm)	H (at.%)	C (at.%)	W (at.%)	Ar (at.%)	O (at.%)	Density (g/cm^3)
10	26.4 ± 1.3	65.6 ± 3.3	5.0 ±0.25	1.58 ± 0.08	1.40 ± 0.07	3.9 ± 0.20
15	26.6 ± 1.3	65.8 ± 3.3	4.6 ±0.23	1.50 ± 0.08	1.50 ± 0.08	3.6 ± 0.18
20	28.3 ± 1.4	64.9 ± 3.2	4.3 ±0.22	1.00 ± 0.05	1.50 ± 0.08	3.2 ± 0.16
25	29.1 ± 1.5	65.5 ± 3.3	4.0 ± 0.20	0.45 ± 0.02	0.90 ± 0.05	3.4 ± 0.17
30	30.8 ± 1.5	64.3 ± 3.2	3.5 ± 0.13	0.20 ± 0.01	1.30 ± 0.07	3.0 ± 0.15
40	30.5 ± 1.5	65.1 ± 3.3	2.9 ± 0.15	0.04 ± 0.00	1.40 ± 0.07	2.1 ± 0.11

The decrease in the W content with the increase in the CH_4 flow could be mainly attributed to two main factors: (i) for similar sputtering conditions, the number of C atoms reaching the substrate increased (from the reactive gas CH_4) and thus the relative number of W atoms decreased; and (ii) depositing in reactive conditions may promote poisoning of the W pellets of the target with the consequent decrease in the sputtering rate of W. The same trend was observed for the Ar content in the coatings which, in this case, was probably due to the lower strain in the films, which provided a lower sticking probability of the noble atoms with almost zero incorporation for the least-dense films. The oxygen content in the coatings was rather low (<2 at%) and appeared independent of the CH_4 flow. It originated from the residual oxygen in the vacuum chamber and/or from atmospheric contamination after deposition. Since the samples were stored in the same conditions, it was expected that the contamination from air exposure would be the same for all samples. Thus, the small differences in the O content in the coatings were probably related to the base vacuum before deposition.

The H in the coatings was from the CH_4 gas; therefore, the IBA showed that the H content increased from about 26 at% to 31 at% with the CH_4 flow rate. The corresponding decrease in the coating density was from 3.9 to 2.1 g/cm^3. This decrease must be understood not only by the decreasing average atomic weight of the coating, but also by the increasing average bond length. Both the composition and the structure of the coatings significantly influenced the density.

The influence of the H content on the bonding structure of the coatings was analyzed with Raman spectroscopy. Figure 1 presents the Raman spectra from 200 to 2000 cm^{-1} for all the coatings. Three main parameters were considered in the analysis: the G peak position; the D over G intensities ratio, ID/IG; and the FWHM (G). Moreover, the photoluminescence background (PL) was also analyzed, which can give a signature for the H content. For higher H contents (<~40 at%), the literature shows that the background overshadows the Raman signal of the coatings [9,15,16,24]. To determine the H content in DLC coatings, Casiraghi et al. [9] established a simple formula that relates the H content with the photoluminescence (PL) signal. The values of PL (m/I(G)) were calculated from the ratio between the slope (m)(background of the spectra in the range 1050 to 1800 cm^{-1}) and the intensity of the G peak IG.

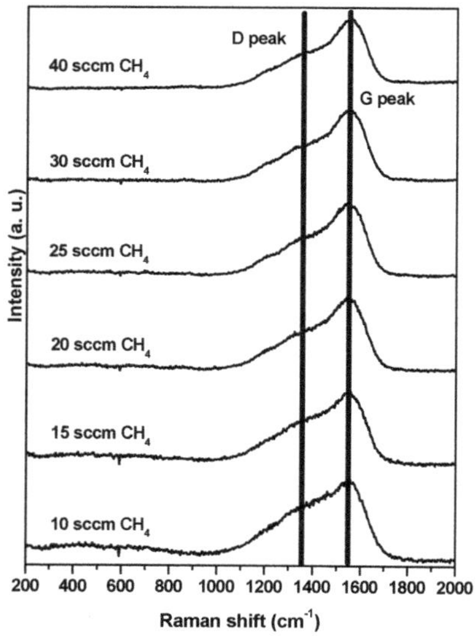

Figure 1. Raman spectra for all samples. For each sample, two measurements were acquired, and the results in the figure are the average of both.

Figure 2 presents the evolution of the m/IG values with the methane flow rate. In addition to the H content, there are other factors that can influence the values of m/IG, including the type of structure of the coatings [25], contamination due to poor vacuum conditions during the deposition or water absorbed after exposure to air [26,27] and alloying coatings with a metal [28]. With an increasing CH_4 flow rate, m/IG also increased (see Figure 2), suggesting an increase in the H content in the films. The slightly different evolution from the one derived by Casiraghi et al. [9] can be attributed either to the presence of W in the coatings or the differences in the structure related to the conditions in the deposition process. For chromium-alloyed DLC coatings (with a similar H content), it was observed that the decrease in the metal content gave an increase in the slope of the Raman background [29]. The coatings deposited in this work presented two features that contributed to the increase in the m/I(G) with the increase in the CH_4 flow, which were (i) the increase in the H content and (ii) the decrease in the W content as the CH_4 flow increased.

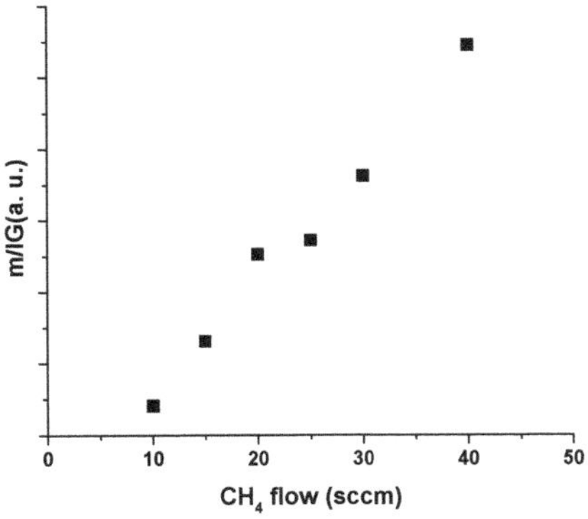

Figure 2. Evolution of m/IG with methane flow rate for the a-C(W):H coatings.

Figure 3a presents the G position and FWHM (G), and Figure 3b presents the ID/IG ratio as functions of the CH_4 flow. The G peak position shifted to a lower wave number with the increasing H content, reaching a minimum value of 25 sccm of CH_4; at this value, an inversion in the trend was observed. The FWHM (G) and ID/IG ratio decreased from 148 to 138 cm^{-1} and from 1.05 to 0.7, respectively, as the CH_4 flow increased. Casiraghi et al. [24] showed that the evolution of the Raman parameters was different for H content higher or lower than a particular value: in their case, it was 20 at%, and in our case, it was a little higher. For amorphous carbon coatings, ID/IG is related to the size of the sp^2 phase organized in rings [15]; thus, a low ID/IG means that the molecular structure is mainly organized in chains even though rings can also be present, but with the π bonds not fully delocalized. FWHM(G) is a measure of the disorder, particularly that arising from bond angle and bond length distortions. A structure organized more in chains (with a polymer-like character) should give rise to a narrowing of the G band, as seen in Figure 3a.

(a) (b)

Figure 3. Raman analysis of the coatings deposited with different CH_4 flows: (**a**) G position and FWHM (G) and (**b**) ID/IG.

In relation to the G peak position, according to Ferrari and Robertson [16], a shift in the G peak to lower wave numbers is related to bond disorder while the inverse is related to chain formation or clustering. These factors act as competing forces on the shape of the Raman spectra.

The adhesion of the coatings was evaluated using a conventional scratch test. The lowest value for the adhesive failure of the coatings was 42 N, and from the six coatings deposited, only two showed adhesive failures with similar values. Those failures were in the end of the scratch, and no extensive failures covering all areas of the scratch were observed. Figure 4 presents the two type of behaviors observed for all coatings. No clear relations of the adhesive failures to the increase in the CH_4 flow was found. The coatings were deposited with increasing flows from 10 to 40 sccm of CH_4, and only the ones with 15 and 20 sccm of CH_4 flows presented adhesive failures.

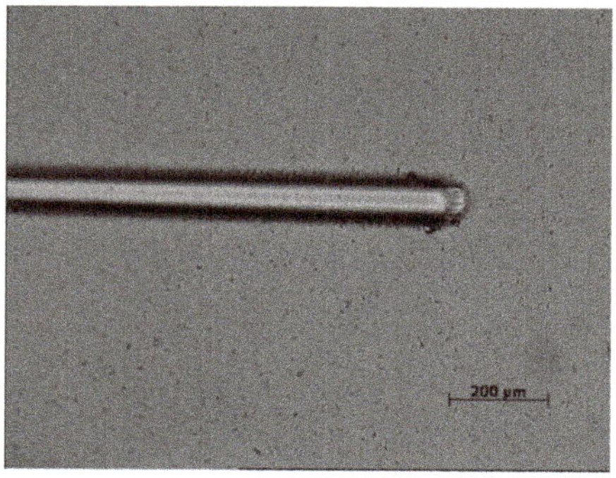

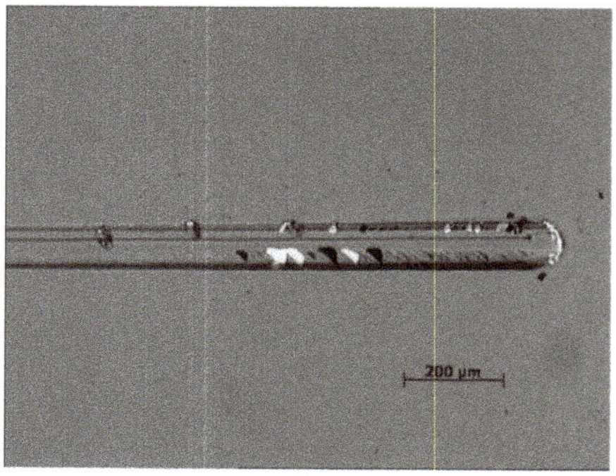

Figure 4. Optical micrographs of the scratch test: (**a**) coating deposited on steel substrates with 40 sccm of CH_4, (**b**) coating deposited with 20 sccm of CH_4.

The mechanical properties of the coatings were evaluated with nanoindentation. The hardness and Young's modulus had similar behavior with the increase in theCH_4 flow, as both showed a decreasing trend, as shown in Figure 5. Although there was a decrease in the W content from 5 to 2.9 at% (Table 1), this behavior was mainly related to the incorporation of H in the coating due to the changes that it caused to the bonding network and structure. With the increasing H content, the mechanical properties of the coatings tended to be determined by the polymer-like structure due to more C-H sp^3 hybridized bonds.

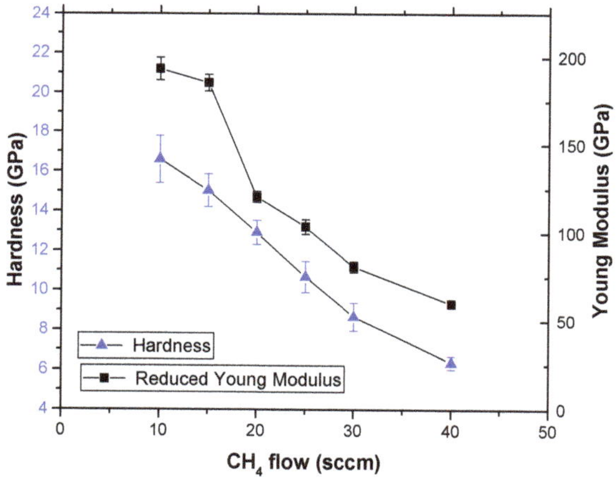

Figure 5. Evolution of the hardness and Young's modulus with the increase of the CH_4 flow.

Figure 6 presents the typical nanoindentations curves of three coatings deposited with increasing CH_4 flow rates. The mechanical work is divided into two parts: the plastic deformation *Wp* and the elastic recovery *We*. From the indentation curves, it can be observed that with the increase in the H content, the hardness decreased as expected and the elastic recovery increased. This behavior was also observed by other authors [18] and might be related to the continuous increase in the number of sp^3 hybridized bonds in the film, as corroborated by the Raman analysis.

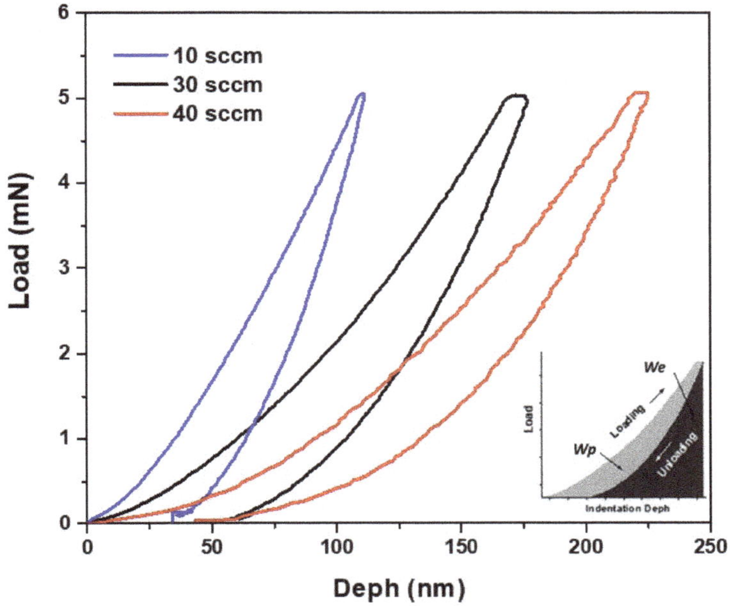

Figure 6. Typical indentation curves for coatings deposited with three different flows of CH_4.

The tribological characterization of the coatings was performed via a pin-on-disc at room conditions with a normal load of 10 N against a 100Cr6 6 mm steel ball. Figure 7a presents the evolution of the friction during the test for the different compositions studied in this work. Two distinct behaviors of the friction can be observed; coatings deposited with up to 20 sccm of CH_4 presented significant fluctuations of the friction during the whole test, while coatings deposited with 25 sccm of CH_4 or higher flows presented a more stable friction during the test.

This behavior can be associated with the formation of a well-adherent homogeneous transfer layer on the sliding ball, as was observed (Figure 8b). For the coatings with lower H content (10 sccm of CH_4), such a transfer layer was not observed (Figure 8a). The increase in the H content promoted a decrease in the average friction (Figure 7b), a behavior usually observed for DLC coatings deposited with increasing H content [30]. The formation of the transfer layer, usually attributed to the graphitization of the coating, can explain the decrease in the friction. Concerning the wear rate for the coating and the sliding partner (100Cr6 steel ball), a gradual decrease was observed with the increase in the H content in spite of their lower hardness. There were several factors that could have influenced the tribological behavior for this set of coatings. One that could have had a major role in the tribological performance was the H content in the coatings. Additionally, the mechanical properties could have influenced the results. Finally, the types of bonds in the coating also contributed to the performance observed for the different coatings tested.

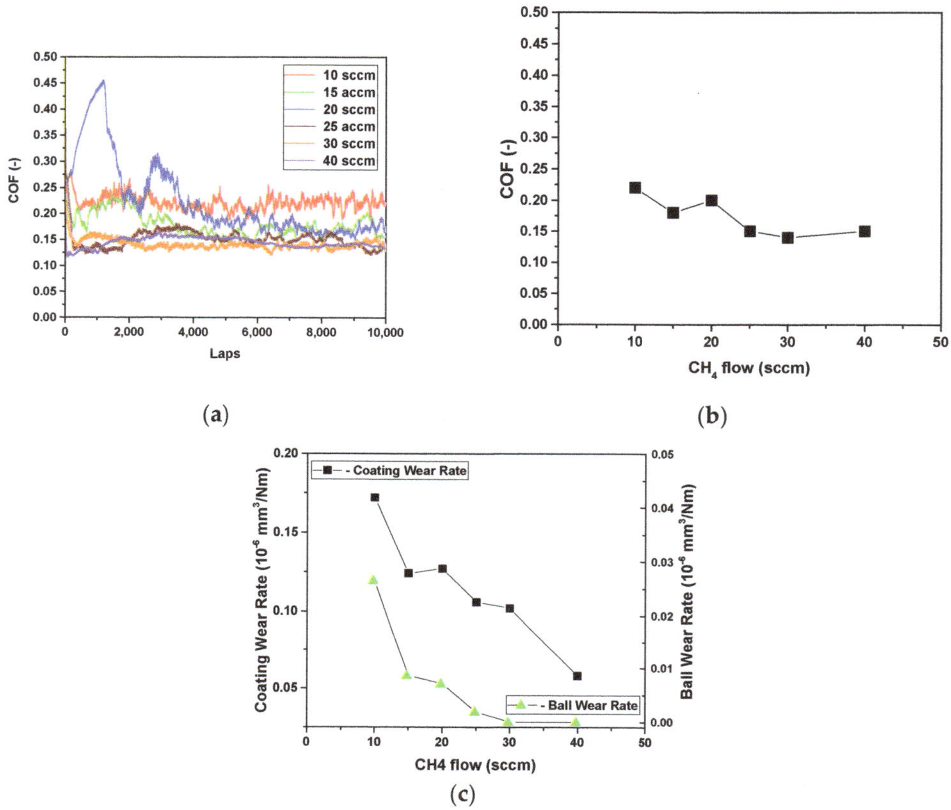

Figure 7. Tribological characterization of the coatings: (**a**) coefficient of friction (COF) plots for all coatings, (**b**) evolution of the average COF with the increase in the CH_4 flow, (**c**) coating and ball-wear rate with the increase in the CH_4 flow.

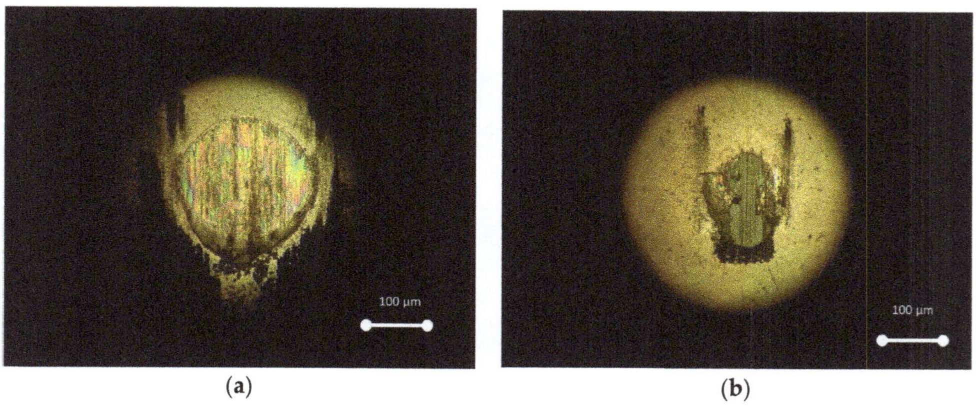

Figure 8. Optical micrograph of the sliding ball after the test with coating deposited with (**a**) 10 sccm of CH_4, (**b**) 40 sccm of CH_4.

4. Conclusions

A set of coatings with an increasing CH_4 flow were successfully deposited on steel and silicon substrates. The IBA analysis showed that the coatings deposited with increasing CH_4 flow rates had markedly lower W and Ar content, whereas the H content increased somewhat. However, the film density reduced markedly with the higher H content. For the same CH_4 variation, Raman spectroscopy showed that the structure of the coatings had a more polymer-like character. This evolution of the structure gave rise both to the decrease in both the hardness and Young's modulus and also to the decrease in the film density. Therefore, it is possible to have a qualitative estimation of the H content and the mechanical properties of the coatings when one knows the flows of CH_4 and uses Raman spectroscopy.

The increase in the H content in the coatings did not significantly influence the coating adhesion. However, the tribological behavior was influenced by the increase in the H content, with a decrease in the friction and wear of both the coating and the sliding partner due to the formation of a well-adherent transfer layer on the latter. The formation of the adherent transfer layer was facilitated with the increase in the H content, mainly due to the fact that there was a decrease in the hardness and Young's modulus as the H content in the coating increased. Thus, this will contribute to the decrease in the wear and friction by the formation of a transfer layer that protects both of the sliding partners.

Supplementary Materials: The following supporting information can be downloaded at: https://www.mdpi.com/article/10.3390/coatings13010092/s1, Table S1: Misfit factor for H signal; Table S2: C/H ratio for all samples, from fitted data. References [31–44] are cited in the supplementary materials.

Author Contributions: Conceptualization, methodology, coatings deposition and writing—original draft preparation, M.E.; coatings characterization; M.E. and C.J.; writing—review and editing, F.F. and A.C. All authors have read and agreed to the published version of the manuscript.

Funding: This research was funded by CEMMPRE–UIDB/ 00285/2020 (cofinanced by FEDER and FCT-Fundação para a Ciência e a Tecnologia (COMPETE)). The IBA was funded by the European Community under the program «Support of Public and Industrial Research Using Ion Beam Technology (SPIRIT)» (EC contract no. 227012).

Institutional Review Board Statement: Not applicable.

Informed Consent Statement: Not applicable.

Data Availability Statement: Not applicable.

Conflicts of Interest: The authors declare no conflict of interest. The funders had no role in the design of the study; in the collection, analyses or interpretation of data; in the writing of the manuscript; or in the decision to publish the results.

References

1. Matthews, A.; Eskildsen, S.S. Engineering applications for diamond-like carbon. *Diam. Relat. Mater.* **1994**, *3*, 902–911. [CrossRef]
2. Czyzniewski, A. Optimising deposition parameters of W-DLC coatings for tool materials of high speed steel and cemented carbide. *Vacuum* **2012**, *86*, 2140–2147. [CrossRef]
3. Czyzniewski, A. The effect of air humidity on tribological behaviours of W–C:H coatings with different tungsten contents sliding against bearing steel. *Wear* **2012**, *296*, 547–557. [CrossRef]
4. Voevodin, A.A.; Prasad, S.V.; Zabinski, J.S. Nanocrystalline carbide/amorphous carbon composites. *J. Appl. Phys.* **1997**, *82*, 855–858. [CrossRef]
5. Corbella, C.; Oncins, G.; Gómez, M.; Polo, M.; Pascual, E.; García-Céspedes, J.; Andújar, J.; Bertran, E. Structure of diamond-like carbon films containing transition metals deposited by reactive magnetron sputtering. *Diam. Relat. Mater.* **2005**, *14*, 1103–1107. [CrossRef]
6. Nilsson, D.; Svahn, F.; Wiklund, U.; Hogmark, S. Low-friction carbon-rich carbide coatings deposited by co-sputtering. *Wear* **2003**, *254*, 1084–1091. [CrossRef]
7. Meng, W.J.; Gillispie, B.A. Mechanical properties of Ti-containing and W-containing diamond-like carbon coatings. *J. Appl. Phys.* **1998**, *84*, 4314–4321. [CrossRef]
8. Suzuki, M.; Ohana, T.; Tanaka, A. Tribological properties of DLC films with different hydrogen contents in water environment. *Diam. Relat. Mater.* **2004**, *13*, 2216–2220. [CrossRef]

9. Casiraghi, C.; Piazza, F.; Ferrari, A.; Grambole, D.; Robertson, J. Bonding in hydrogenated diamond-like carbon by Raman spectroscopy. *Diam. Relat. Mater.* **2005**, *14*, 1098–1102. [CrossRef]
10. Lofaj, F.; Kabátová, M.; Kvetková, L.; Dobrovodský, J. The effects of deposition conditions on hydrogenation, hardness and elastic modulus of W-C:H coatings. *J. Eur. Ceram. Soc.* **2020**, *40*, 2721–2730. [CrossRef]
11. Lofaj, F.; Kabátová, M.; Dobrovodský, J.; Cempura, G. Hydrogenation and hybridization in hard W-C:H coatings prepared by hybrid PVD-PECVD method with methane and acetylene. *Int. J. Refract. Met. Hard Mater.* **2020**, *88*, 105211. [CrossRef]
12. Libardi, J.; Grigorov, K.; Massi, M.; Otani, C.; Ravagnani, S.; Maciel, H.; Guerino, M.; Ocampo, J. Comparative studies of the feed gas composition effects on the characteristics of DLC films deposited by magnetron sputtering. *Thin Solid Films* **2004**, *459*, 282–285. [CrossRef]
13. Robertson, J. Requirements of ultrathin carbon coatings for magnetic storage technology. *Tribol. Int.* **2003**, *36*, 405–415. [CrossRef]
14. Kahn, M.; Paskvale, S.; Čekada, M.; Schöberl, T.; Waldhauser, W.; Mitterer, C.; Pelicon, P.; Brandstätter, E. The relationship between structure and mechanical properties of hydrogenated amorphous carbon films. *Diam. Relat. Mater.* **2010**, *19*, 1245–1248. [CrossRef]
15. Ferrari, A.C.; Robertson, J. Interpretation of Raman spectra of disordered and amorphous carbon. *Phys. Rev. B* **2000**, *61*, 14095. [CrossRef]
16. Ferrari, A.C.; Robertson, J. Resonant Raman spectroscopy of disordered, amorphous, and diamondlike carbon. *Phys. Rev. B* **2001**, *64*, 075414. [CrossRef]
17. Robertson, J. Daimond–Like amorphous carbon. *Mat. Sci. Eng.* **2002**, *R37*, 129. [CrossRef]
18. Dai, W.; Wu, G.; Wang, A. Structure and elastic recovery of Cr–C:H films deposited by a reactive magnetron sputtering technique. *Appl. Surf. Sci.* **2010**, *257*, 244–248. [CrossRef]
19. Jeynes, C.; Colaux, J.L. Thin film depth profiling by ion beam analysis. *Analyst* **2016**, *141*, 5944–5985. [CrossRef]
20. Fernandes, J.; Trindade, A.; Menezes, L.; Cavaleiro, A. A model for coated surface hardness. *Surf. Coat. Technol.* **2000**, *131*, 457–461. [CrossRef]
21. Gurbich, A. Evaluated differential cross-sections for IBA. *Nucl. Instrum. Methods B* **2010**, *268*, 1703–1710. [CrossRef]
22. Jeynes, C.; Palitsin, V.; Grime, G.; Pascual-Izarra, C.; Taborda, A.; Reis, M.; Barradas, N. External beam Total-IBA using DataFurnace. *Nucl. Instrum. Methods B* **2020**, *481*, 47–61. [CrossRef]
23. Barradas, N.P.; Jeynes, C. Advanced physics and algorithms in the IBA DataFurnace. *Nucl. Instruments Methods B* **2008**, *266*, 1875–1879. [CrossRef]
24. Casiraghi, C.; Ferrari, A.C.; Robertson, J. Raman spectroscopy of hydrogenated amorphous carbons. *Phys. Rev. B* **2005**, *72*, 085401. [CrossRef]
25. Henley, S.J.; Carey, J.D.; Silva, S.R.P. Room temperature photoluminescence from nanostructured amorphous carbon. *Appl. Phys. Lett.* **2004**, *85*, 6236–6238. [CrossRef]
26. Rodil, S.E.; Muhl, S.; Maca, S.; Ferrari, A.C. Optical gap in carbon nitride films. *Thin Solid Film* **2003**, *433*, 119. [CrossRef]
27. Fanchini, G.; Ray, S.; Tagliaferro, A. Photoluminescence investigation of carbon nitride-based films deposited by reactive sputtering. *Diam. Relat. Mater.* **2003**, *12*, 1084–1087. [CrossRef]
28. Zhang, H.; Wu, W.; Gong, C.; Wang, W.; He, Z.; Li, J.; Ju, X.; Tang, Y.; Xie, E. Structural and optical properties of Fe-doped hydrogenated amorphous carbon films prepared from trans-2-butene by plasma enhanced metal organic chemical vapor deposition. *Appl. Phys. A* **2010**, *98*, 895–900. [CrossRef]
29. Pardo, A.; Gómez-Aleixandre, C.; Orwa, J.; Cimmino, A.; Prawer, S. Modification of characteristics of diamond-like carbon thin films by low chromium content addition. *Diam. Relat. Mater.* **2012**, *26*, 39–44. [CrossRef]
30. Lugo, D.; Silva, P.; Ramirez, M.; Pillaca, E.; Rodrigues, C.; Fukumasu, N.; Corat, E.; Tabacniks, M.; Trava-Airoldi, V. Characterization and tribologic study in high vacuum of hydrogenated DLC films deposited using pulsed DC PECVD system for space applications. *Surf. Coat. Technol.* **2017**, *332*, 135–141. [CrossRef]
31. Boudreault, G.; Elliman, R.G.; Grötzschel, R.; Gujrathi, S.C.; Jeynes, C.; Lennard, W.N.; Rauhala, E.; Sajavaara, T.; Timmers, H.; Wang, Y.Q.; et al. Round Robin: Measurement of H implantation distributions in Si by elastic recoil detection. *Nucl. Instrum. Methods B* **2004**, *222*, 547–566. [CrossRef]
32. Abriola, D.; Barradas, N.P.; Bogdanović-Radović, I.; Chiari, M.; Gurbich, A.F.; Jeynes, C.; Kokkoris, M.; Mayer, M.; Ramos, A.R.; Shi, L.; et al. Development of a reference database for Ion Beam Analysis and future perspectives. *Nucl. Instrum. Methods B* **2011**, *269*, 2972–2978. [CrossRef]
33. Ziegler, J.F. SRIM-2003. *Nucl. Instrum. Methods B* **2004**, *219*, 1027–1036. [CrossRef]
34. Wielopolski, L.; Gardner, R.P. Prediction of the pulse-height spectral distortion caused by the peak pile-up effect. *Nucl. Instrum. Methods B* **1976**, *133*, 303–309. [CrossRef]
35. Molodtsov, S.L.; Gurbich, A.F. Simulation of the pulse pile-up effect on the pulse-height spectrum. *Nucl. Instrum. Methods B* **2009**, *267*, 3484–3487. [CrossRef]
36. Pascual-Izarra, C.; Barradas, N.P. Introducing routine pulse height defect corrections in IBA. *Nucl. Instrum. Methods B* **2008**, *266*, 266–270. [CrossRef]
37. Barradas, N.P. Rutherford backscattering analysis of thin films and superlattices with roughness. *J. Phys. D Appl. Phys.* **2001**, *34*, 2109–2116. [CrossRef]

38. Simon, A.; Jeynes, C.; Webb, R.P.; Finnis, R.; Tabatabaian, Z.; Sellin, P.J.; Breese, M.B.H.; Fellows, D.F.; van den Broek, R.; Gwilliam, R.M. The new Surrey ion beam analysis facility. *Nucl. Instrum. Methods B* **2004**, *219*, 405–409. [CrossRef]
39. Grime, G.W.; Dawson, M. Recent developments in data acquisition and processing on the Oxford scanning proton microprobe. *Nucl. Instrum. Methods B* **1995**, *104*, 107–113. [CrossRef]
40. Blaauw, M.; Campbell, J.L.; Fazinić, S.; Jakšić, M.; Orlic, I.; van Espen, P. The 2000 IAEA intercomparison of PIXE spectrum analysis software. *Nucl. Instrum. Methods B* **2002**, *189*, 113–122. [CrossRef]
41. Jeynes, C.; Barradas, N.P.; Marriott, P.K.; Boudreault, G.; Jenkin, M.; Wendler, E.; Webb, R.P. Elemental thin film depth profiles by ion beam analysis using simulated annealing-a new tool. *J. Phys. D* **2003**, *36*, R97. [CrossRef]
42. Boudreault, G.; Jeynes, C.; Wendler, E.; Nejim, A.; Webb, R.P.; Wätjen, U. Accurate RBS measurement of ion implant doses in silicon. *Surf. Interface Anal.* **2002**, *33*, 478–486. [CrossRef]
43. Barradas, N.P.; Arstila, K.; Battistig, G.; Bianconi, M.; Dytlewski, N.; Jeynes, C.; Kótai, E.; Lulli, G.; Mayer, M.; Rauhala, E.; et al. Summary of "IAEA intercomparison of IBA software". *Nucl. Instrum. Methods B* **2008**, *266*, 1338–1342. [CrossRef]
44. Pascual-Izarra, C.; Barradas, N.P.; Reis, M.A.; Jeynes, C.; Menu, M.; Lavedrine, B.; Ezrati, J.J.; Röhrs, S. Towards truly simultaneous PIXE and RBS analysis of layered objects in cultural heritage. *Nucl. Instrum. Methods B* **2007**, *261*, 426–429. [CrossRef]

Disclaimer/Publisher's Note: The statements, opinions and data contained in all publications are solely those of the individual author(s) and contributor(s) and not of MDPI and/or the editor(s). MDPI and/or the editor(s) disclaim responsibility for any injury to people or property resulting from any ideas, methods, instructions or products referred to in the content.

coatings

MDPI

Article

Effect of Standoff Distance on Corrosion Resistance of Cold Sprayed Titanium Coatings

Mieczyslaw Scendo * and Katarzyna Staszewska-Samson

Institute of Chemistry, Jan Kochanowski University in Kielce, Uniwersytecka 7, PL-25406 Kielce, Poland
* Correspondence: scendo@ujk.edu.pl

Abstract: A titanium protective layer was deposited onto Al7075 substrate (Al7075/Ti) by a cold spray method (CS) with different standoff distances (SoD) of the nozzle from the specimen surface. The aim of this research was to study the influence on the mechanical properties and corrosion resistance of the Ti coating on Al7075 substrate. The surface and microstructure of Al7075/Ti was observed by a scanning electron microscope (SEM). The corrosion test of the materials was carried out by using the electrochemical method. The SoD had a significant effect on the microstructure of the coatings and their adhesion onto Al7075 substrate. The highest level of microhardness (248 HV0.3) value was achieved for deposits obtained with the SoD of 70 mm. The corrosion tests showed that the mechanism of electrochemical corrosion of titanium coatings is a multi-stage process, and the main product of the corrosion process was $(TiO_2)_{ads}$. However, the polarization resistance (R_p = 49 kΩ cm^2) of the Al7075/Ti coatings was the highest, while the corrosion rate (υ_{corr} = 13.90 mm y^{-1}) was the lowest, for SoD of 70 mm.

Keywords: Al7075 substrate; standoff distance; cold spray; titanium coating; microstructure; corrosion rate

1. Introduction

Cold spraying (CS) is a coating technology based on aerodynamics and high-speed impact dynamics. In this process, spray particles (5–50 μm) are accelerated to a high velocity (300–1200 m/s) by a high-speed gas flow that is generated through a convergent–divergent de Laval type nozzle [1–3]. Pressurized gas (N_2 or He) is heated, typically by electric energy, to temperatures in the range of 300 °C to 800 °C and then directed to a nozzle to produce a supersonic inert gas stream [4]. Obtaining supersonic velocity by spray particles is a prerequisite for obtaining a coating that adheres well to the substrate, which ensures that it obtains the appropriate mechanical properties [5,6]. When the particles exit the nozzle and impinge on the target surface, they undergo significant plastic deformations resulting from collisions and bond to the substrate. Thus, there is a clear change in particle size [7]. A coating is formed through the intensive plastic deformation of particles impacting on a substrate at a temperature well below the melting point of the spray material. The main result is that CS can minimize effects of oxidation, melting, evaporation and other common problems suffered in thermal spraying [8]. Therefore, CS is used for the production and repair of metal coatings to increase mechanical properties and improve corrosion resistance of various metal components. However, it has been found that the particle size of the powder used to produce the coating has little effect on the deformation of the entire particle as it hits the substrate at high speed. It turned out that the particle flattening coefficient increases very markedly with the increase in the velocity of impact of the metal powder particles on the substrate surface. In contrast, the temperature at the localized contact surfaces increases significantly due to the possible adiabatic shearing process. The critical speed for particle deposition could be estimated with appropriate material properties [9,10]. The distance between the nozzle and the surface to which a new coating is applied has a significant impact on the mechanical properties of metallic coatings. On the other hand, for cold spray deposition, increasing of the standoff distance (SoD) was generally thought to result in a lower particle speed due to the drag force exerted on the particle. Therefore, the protective coatings obtained did not have high mechanical parameters. On the other hand, too small a distance of the nozzle from the surface of the substrate results in lower quality coatings with low mechanical parameters. Therefore, a small SoD will reduce the particle velocity and deposition efficiency, which slows down the particle velocity before impact [11–19]. In order for the layer to be effectively formed on the surface of the substrate, the metal particles to be deposited must move at a supersonic speed clearly greater than the critical speed. It is worth adding that if the metal particle velocity is lower than the critical speed, the substrate may be significantly damaged as a result of its abrasion. Importantly, by increasing the powder feed rate in the nozzle, the velocity of the deposited metal particles is reduced due to gas–particle interactions as the particles move away from the nozzle outlet [20–22]. On the other hand, the production of metallic coatings is a multi-step process. Initially, a thin layer of material is deposited on the substrate. This stage is characterized by the direct interaction of the particles with the substrate and largely depends on the degree of preparation and the properties of the substrate. Subsequent layers are formed as a result of multiple overlapping coating material particles [23].

The aluminium–zinc–magnesium (i.e., Al7075) alloys have a greater response to heat treatment than the binary aluminium–zinc alloys, resulting in higher strengths. The additions of zinc and magnesium, however, decrease the corrosion resistance. Thus, the alloy Al7075 must be protected against corrosion, most often by means of metallic coatings resistant to corrosion. A possible approach to increase the corrosion resistance of this alloy is the deposition of a pure titanium coated layer. The titanium standard potential is ${E_{Ti}}^{2+}{}_{/Ti} = -1.75$ V, so titanium is a non-noble metal. Titanium is a metal with high hardness and mechanical strength. Moreover, under natural conditions, the Ti surface undergoes a passivation, and therefore titanium is classified as a corrosion-resistant metal. However, the lower content of oxides is due to the cold spray being carried out at much lower temperature, and the reaction between the metal and the oxygen in the ambient spray can be greatly reduced or eliminated [24].

There is no reliable information in the literature so far concerning the influence of the distance between the nozzle and the substrate surface on the mechanical and anti-corrosion properties of metal coatings produced by the cold spray method.

The aim of this research was to check the impact influence of the different standoff distances of the nozzle from the specimen surface on the microhardness and corrosion resistance of the Ti coating onto Al7075 substrate. The titanium protective layer was produced by the cold spray method. The corrosion test was carried out by electrochemical method in an acidic chloride solution.

2. Experimental Details

2.1. *Materials and Methods*

The titanium coatings were deposited onto Al7075 substrate (Al7075/Ti) by the cold spray method (CS). Pure titanium powder (99.8 wt.% Ti) was used as feedstock. This powder was manufactured using the hydrite–dehydrite process and supplied by Kamb Import-Export (Warsaw, Poland). On the basis of the analysis of the powder's particle size distribution, it was found that was characterized by the d10 = 18.0 μm, d50 = 35.0 μm, and d90 = 60.0 μm [25], Figure 1.

The cold spray deposition was performed with an Impact Innovations 5/8 System (Impact-Innovations GmbH, Rattenkirchen, Germany). Nitrogen was applied as the process gas to the deposit titanium coatings. The maximum parameters used were a temperature of 800 °C and a pressure of nitrogen of 40 bar [25]. The cold spray equipment is shown in Figure 2.

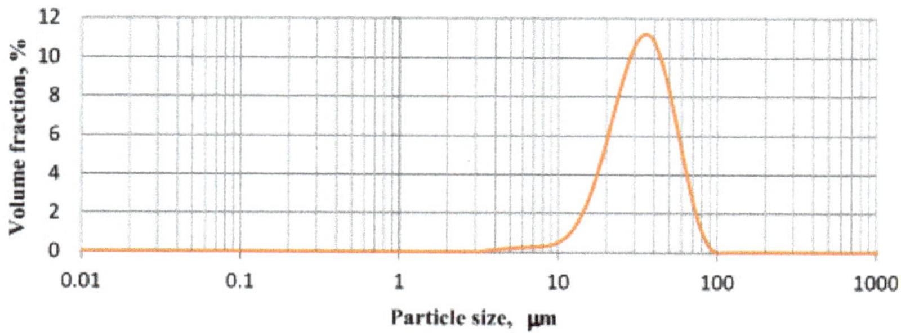

Figure 1. Particle size distribution of titanium powder [25].

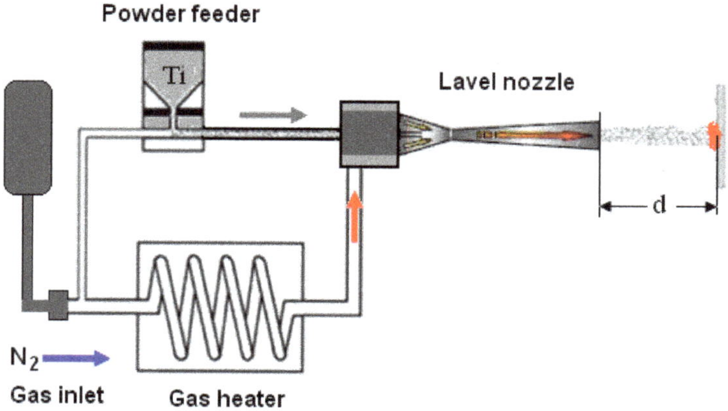

Figure 2. Typical cold spray equipment with a marked distance of the nozzle from the sample surface.

Before applying the Ti coating, the surface of the substrate, i.e., Al7075 was mechanically cleaned with 600, 1200, and 2000 grit sandpaper. Coatings were sprayed maintaining the distance (d) of the nozzle from the specimen surface in the range from 20 to 100 mm, which was systematically increased by 10 mm. The nozzle traverse speed was 400 mm/s. In order to obtain the planned thickness of the titanium coating, i.e., 1.5 mm–2 mm onto Al7075 surface, we made 2 or 3 runs with a cold spray gun [25]. The Al7075/Ti samples were cooled (in about 24 h) to room temperature under a nitrogen atmosphere. The basic parameters of the materials are listed in Table 1.

Table 1. Sample name, and distance of nozzle from sample surface.

Sample	Distance d, mm
Al7075/Ti-20	20
Al7075/Ti-70	70
Al7075/Ti-100	100

Three of the Al7075/Ti samples were selected for the characterization of microstructures and corrosion test (Table 1).

2.2. Solutions

The following reagents were used to make the solutions: FLUKA analytical grade sodium chloride (NaCl) and POCH analytical grade hydrochloric acid (HCl). The concentration of Cl^- ions was 1.2 M, and the pH was 1.5. The electrolyte was not deoxygenated.

2.3. Electrodes

The working electrode (W) was made of Al7075 alloy, which was covered with titanium coating (Al7075/Ti), which was produced by a cold spraying method. The geometric surface area of the W electrode was 1.0 cm^2. The saturated calomel electrode (SCE) was used as the reference. It was connected with the solution using a Luggin capillary. The counter electrode (9 cm^2) was made from platinum foil (99.9% Pt).

2.4. Electrochemical Measurements

The electrochemical experiments were carried out in a conventional three-electrode cell. All electrochemical measurements were made using a potentiostat/galvanostat PG-STAT 128N, (AutoLab, Amsterdam, The Netherlands) with NOVA 1.7 software from the same company.

The potentiodynamic polarization (LSV) curves were recorded. All measurements were carried out under a potential range from −900 mV to −100 mV vs. SCE, whereas the potential change rate was 1 mV s^{-1}, with holding time of 30 s at −900 mV.

The LSV curves were used to designate of the corrosion electrochemical parameters. However, the Stern–Geary equation was used for the calculation of the polarization resistance of the materials. The corrosion rate of materials were appointed using the following equation [26–29]:

$$v_{corr} = 3.268 \times \frac{j_{corr}\,M}{n\,\rho} \tag{1}$$

where j_{corr} is the corrosion current density, M is the molecular weight of the substrate, n is the number of electrons exchanged, and ρ is the density of the material.

The chronoamperometric curves (ChA) were obtained for the potential values which were selected for the characteristic points on the LSV curves.

2.5. Surface Morphologies and Microstructure

The surface morphologies and microstructure were observed by using a scanning electron microscope (SEM, JSM-5400, Joel, Tokyo, Japan). The accelerating voltage of SEM was 20 kV. Prior to the cross-sectional analysis, the coating samples were polished with increasingly fine, (3 µm, 1 µm, and 0.25 µm) diamond suspensions. The X-ray diffraction (XRD) was applied to characterize the phase composition of the coatings using a Bruker D8 Discover diffractometer (Bruker Ltd., Malvern, UK), with Co Kα radiation was of λ = 1.7889 Å.

2.6. Microhardness

The measurement of microhardness of the Al7075/Ti was made by the Vickers method (HV) using a Falcon 500 hardness tester from INNOVATEST (Maastricht, The Netherlands). An indenter was used in the form of a diamond pyramid, whose load varied from 0.02 N to 20 N.

All measurements were carried out at a temperature of 25 ± 0.5 °C which were maintained using an air thermostat.

3. Results and Discussion

3.1. Surface Morphologies

The surface morphologies of the Ti coatings onto Al7075 substrate are presented in Figure 3.

Titanium coatings adhere well onto Al7075 alloy. As shown in Figure 3, along with the change in SoD, a change in the surface structure of the Ti coating onto Al7075 substrate was observed. The most homogeneous and smooth surface of the Ti coating was obtained when the nozzle distance from the sample surface was 70 mm (Figure 3b).

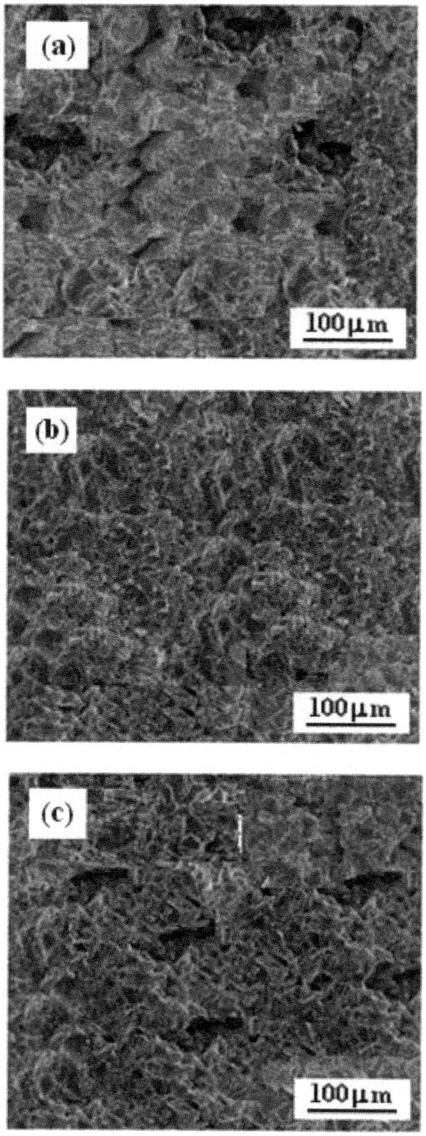

Figure 3. Surface morphologies of cold sprayed titanium coatings onto Al7075 substrate deposited at SoD of: (**a**) 20 mm, (**b**) 70 mm, and (**c**) 100 mm.

Figure 4 presents the X-ray diffraction patterns recorded for the Ti powder and cold sprayed coating deposited onto Al7075 substrate at SoD of 70 mm.

It can be seen that the SoD of 70 mm and high gas temperature of 800 °C do not cause the formation of oxides onto the Al7075/Ti surface. However, similar results were obtained for other samples (Table 1), but they are not cited in this work.

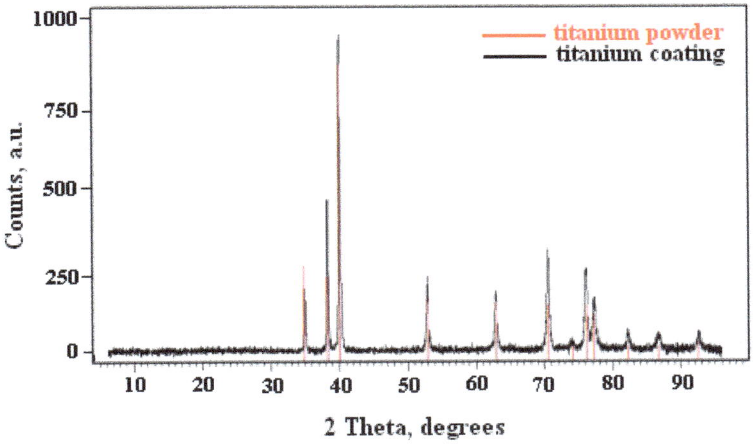

Figure 4. X-ray spectrum of cold sprayed titanium coating onto Al7075 substrate deposited at SoD of 70 mm.

3.2. Microhardness

In Table 2 shows how the SoD affects the microhardness of the Ti coatings onto Al7075 substrate. Similar observations were reported by [25].

Table 2. Microhardness of titanium coatings onto Al7075 substrate.

Sample	Microhardness HV0.3
Al7075/Ti-20	200 ± 2
Al7075/Ti-70	248 ± 1
Al7075/Ti-100	218 ± 3

The highest HV0.3 value was recorded for the titanium coating that was produced when the SoD was 70 mm (Table 2). The measurement results show a large deviation of the HV0.3 value, which is related to the different degree of deformation of the titanium particles on the Al7075 surface.

3.3. Corrosion Test

The corrosion tests of the Al7075/Ti coatings were carried out by the electrochemical method in 1.2 M Cl^- solution (pH 1.5). The potentiodynamic polarization curves (LSV) of the titanium coatings onto the Al7075 alloy, depending on the nozzle distance from the sample surface, are shown in Figure 5.

The cathode branches (Figure 5) correspond to the simplified reduction reaction of hydrogen ions [26–28]:

$$Ti + 2\,H^+ \rightarrow Ti + H_2 - 2\,e^- \quad (2)$$

When the electrode potential is changed in the anode direction, the surface of the Ti electrode was oxidized:

$$Ti \rightarrow Ti^{2+} + 2\,e^- \quad (3)$$

Moreover, subsequently, an adsorbed oxide layer was formed on the surface of the electrode according to the reaction:

$$Ti^{2+} + 2\,H^+ + 3/2O_2 \rightarrow (TiO_2)_{ads} + H_2O + 8\,e^- \quad (4)$$

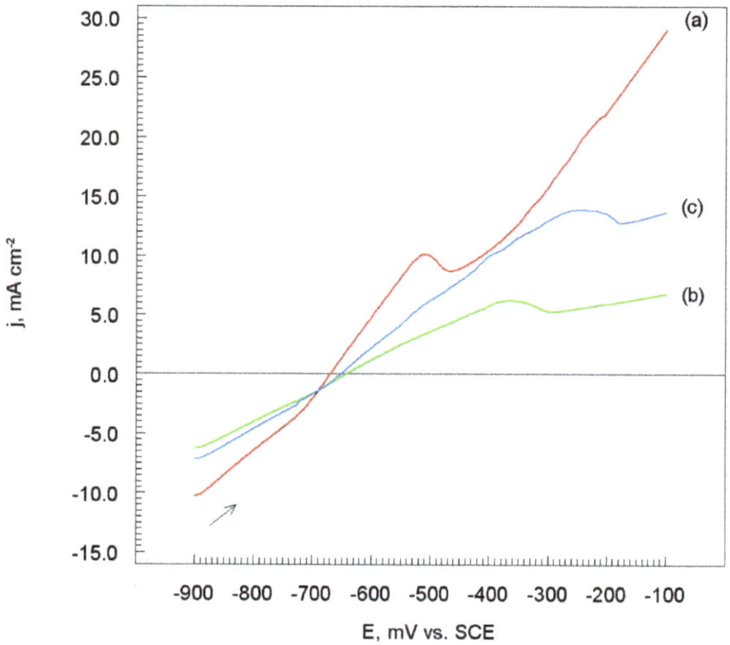

Figure 5. Potentiodynamic polarization curves of titanium coatings onto Al7075 substrate deposited at SoD of: (**a**) 20 mm, (**b**) 70 mm, and (**c**) 100 mm. Solutions contained 1.2 M Cl^-, pH 1.5, dE/dt 1 mV s^{-1}.

The Al7075/Ti surface was covered with a white layer of titanium (IV) oxide. Therefore, on the potentiodynamic polarization curves (Figure 5), peaks are observed for the electrode potentials of: −510 mV (curve (a)), −360 mV (curve (b)), and −230 mV (curve (c)) vs. SCE, respectively. Adsorbed titanium (IV) oxide (Equation (4)) clearly limits the effects of electrochemical corrosion on the titanium surface, especially for Al7075/Ti, which was produced when the nozzle distance from the sample surface was 70 mm (Figure 5, curve (b)). In this case, the protective oxide layer (reaction (4)) is tight and adheres well to the substrate, protecting the Al7075 substrate against contact with a corrosive chloride environment.

Under the conditions of the experiment in the acidic environment, the adsorbed oxide layer dissolves according to a simple chemical reaction:

$$(TiO_2)_{ads} + 4H^+ \rightarrow Ti^{4+} + 2H_2O \quad (5)$$

However, for more positive electrode potentials, the current density increases (Figure 5), hence a further dissolution of the titanium surfaces that have been deposited onto Al7075 substrate.

3.3.1. Corrosion Electrochemical Parameters

The LSV curves (Figure 5) were used to designate the corrosion electrochemical parameters of the Al7075/Ti coatings, i.e., corrosion potential (E_{corr}), corrosion current density (j_{corr}), and cathodic (b_c) and anodic (b_a) Tafel slopes (Figure 6).

The corrosion electrochemical parameters of the titanium coatings onto Al7075 substrate in aggressive chloride solution are listed in Table 3.

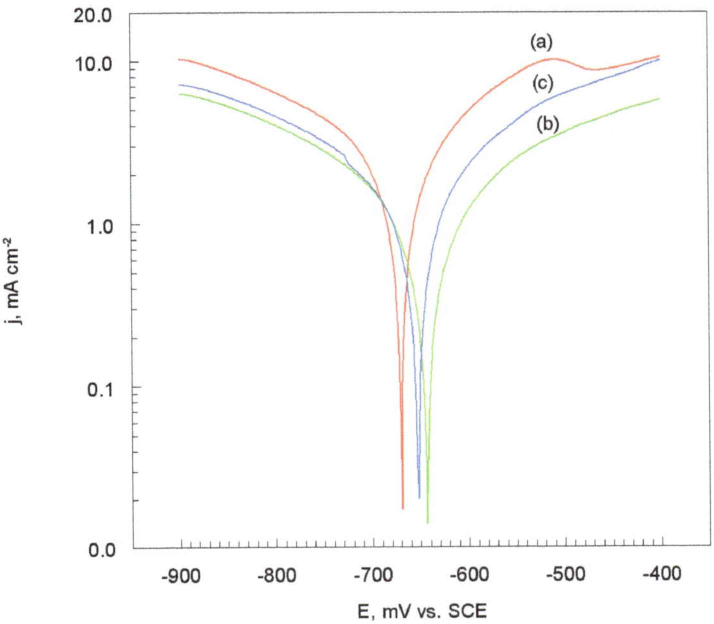

Figure 6. Potentiodynamic polarization curves on a semi-logarithmic (Tafel scale) of titanium coatings onto Al7075 substrate deposited at SoD of: (**a**) 20 mm, (**b**) 70 mm, and (**c**) 100 mm. Solutions contained 1.2 M Cl^-, pH 1.5, dE/dt 1 mV s^{-1}.

Table 3. Corrosion electrochemical parameters of titanium coatings onto Al7075 substrate.

Sample	E_{corr} mV vs. SCE	j_{corr} mA cm^{-2}	$-b_c$	b_a
			mV dec^{-1}	
Al7075/Ti-20	−670	2.80	350	250
Al7075/Ti-70	−643	1.60	340	380
Al7075/Ti-100	−652	1.90	380	300

It turned out that with the increase in the distance of the nozzle from the surface of the samples, the values of the corrosion potential shift towards positive values. Moreover, the most positive value of the corrosion potential was noted for the Al7075/Ti-70 coating (Table 3). Thus, if the distance between the nozzle and the surface of the sample was 70 mm, the Ti coating exhibits the best anti-corrosion properties. This is also evidenced by the lowest value of the corrosion current density (i.e., 1.60 mA cm^{-2}). The slopes of the Tafel's cathode fragments of polarization curves are slightly different between each other (Table 3). Therefore, the mechanism of cathodic hydrogen reduction (Equation (2)) is identical for the tested electrodes. However, the Tafel slopes of the anode segments of the polarization curves are different for each electrode material (Table 3). This suggests that the mechanism of electrochemical corrosion (Equations (3)–(5)) of the tested coatings depends on the method of Ti coating production onto Al7075 alloy (in this case, from the distance of the nozzle from the surface of the substrate).

3.3.2. Polarization Resistance and Corrosion Rate

The polarization resistance (R_p) of the titanium coatings were determined on the basis of the slope of the LSV curves (Figure 5). The corrosion rates (v_{corr}) of the materials were calculated on the basis of Equation (1) [27–29]. The R_p and v_{corr} values of the titanium coatings cold sprayed onto Al7075 alloy are collected in Table 4.

Table 4. Polarization resistance and corrosion rate of titanium coatings onto Al7075 substrate.

Sample	R_p kΩ cm^2	v_{corr} mm y^{-1}
Al7075/Ti-20	23	24.30
Al7075/Ti-70	49	13.90
Al7075/Ti-100	38	16.50

The highest in the electrode's polarization resistance value was noted of the Ti coating, which was obtained when the nozzle distance from the surface of the substrate was 70 mm. Moreover, the electrochemical corrosion rate of this coating is also the smallest compared to Ti coatings that were produced by a different nozzle position (Table 4). It seems clear that metallographic structures of titanium coatings onto Al7075 alloy are completely different.

3.3.3. Chronoamperometric Measurements

Figure 7 shows the ChA curves in 1.2 M Cl^- solution (pH 1.5) of the Ti coating onto Al7075 substrate deposited at SoD of 20 mm, 70 mm, and 100 mm. The electrode potential of the anode process was −150 mV vs. SCE, and was selected on basis of the LSV curves (Figure 5). It was found that, in the conditions of the experiment, the current density values depended on the distance of the nozzle from the surface of the substrate (Figure 7). For each nozzle distance from the Al7075 surface, the current density systematically decreases during electrolysis. Thus, the adsorbed oxide layer i.e., TiO_2 (reaction (4)) protects the Al7075 substrate against contact with an aggressive chloride solution.

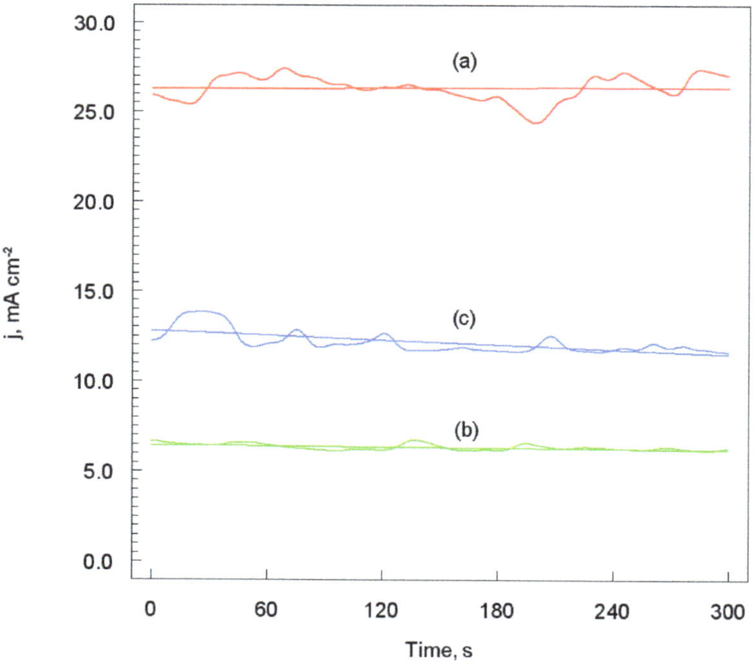

Figure 7. Chronoamperometric curves obtained for −150 mV of titanium coatings onto Al7075 substrate deposited at SoD of: (**a**) 20 mm, (**b**) 70 mm, and (**c**) 100 mm. Solutions contained 1.2 M Cl^-, pH 1.5. (Straight lines refer to the average current density values).

It seems that, under these conditions, the adsorbed oxide layer can be additionally sealed by adsorption of Cl^- ions [29]:

$$MeO + Cl^- + H^+ \rightarrow (MeClOH^-)_{ads} \quad (6)$$

and

$$(MeClOH^-)_{ads} \rightarrow (MeClOH)_{ads} + m\ e^- \quad (7)$$

Unfortunately, the $(MeClOH)_{ads}$ layer in the acidic chloride solution was dissolved in accordance with a chemical reaction:

$$(MeClOH)_{ads} + H^+ \rightarrow Me^{n+} + Cl^- + H_2O \quad (8)$$

Thus, a further sharp increase in the current intensity was observed due to the oxidation of the electrode surface (Figure 5).

In addition, a clear fluctuation of the current density was observed (especially for the distance of 20 mm), which were associated with the oxidation of the surface of the test materials (Figure 7, curve (a)). Thus, in this case, the oxide surface is not homogeneous and does not effectively protect the substrate against corrosion.

Figure 8 shows a scheme of the formation of the protective titanium layer onto Al7075 substrate depending on the distance of the nozzle from the substrate surface.

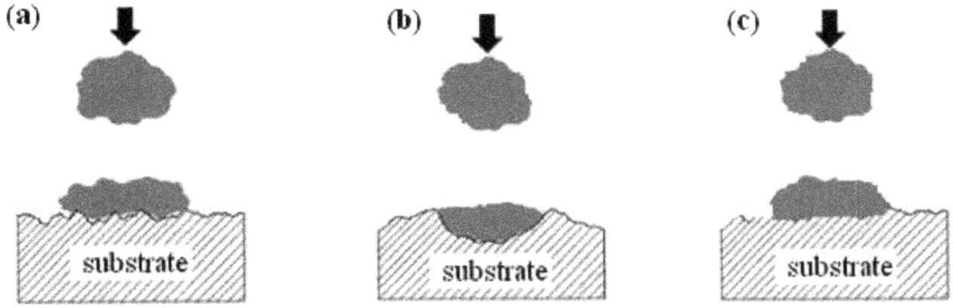

Figure 8. Scheme formation of titanium layer onto Al7075 substrate depending on nozzle distances from sample surface of: (**a**) 20 mm, (**b**) 70 mm, and (**c**) 100 mm.

During the production of the metal layer by the cold spray method, the particle impact speed must exceed the threshold value so that plastic deformation of the metal particles occurs, which causes them to adhere to the surface of the substrate. Moreover, nozzle geometry, powder characteristics and nozzle distance from the sample surface are of fundamental importance with respect to the microstructure, physical, and mechanical properties of the coatings.

If the distance of the nozzle from the substrate was 20 mm, the Ti particles were subjected to excessive deformation, which resulted in a leaky Ti coating onto Al7075 substrate (Figure 8a). It seems that the most compact and tight Ti coating was obtained when the distance of the nozzle from the substrate Al7075 was placed 70 mm. In this case, tightly adhering Ti coating was obtained (Figure 8b). However, during Ti coating production cold spray method when the distance of the nozzle from the substrate was increased to 100 mm, the Ti particles did not deform sufficiently and a leaky porous coating was obtained (Figure 8c). The aggressive chloride electrolyte easily penetrates deep into the structure of the Ti coating, causing corrosion of the substrate.

On the other hand, an important aspect of the powder morphology is that the particle shape of the coating material has a significant influence on the mechanical properties of the newly formed coating. It turns out that irregularly shaped metal particles hitting the substrate cause high internal stresses in the new coating. Ajdelsztajn et al. [30] showed that when the shell metal particles hit the substrate, a localized shear deformation occurs at the particle boundaries. In this way, the shape of the metal particles is changed to be more regular, which favors close contact between the particles and helps to form a metallurgical particle/particle bond. In addition, the irregular morphology of the metal particles will increase the concentration of internal stresses in the coating due to the fact that the load cannot be evenly distributed. Thus, the regular morphology of the coating material particles promotes an even distribution of internal stresses, and thus significantly improves the mechanical properties of the new coatings [28]. Moreover, the concentration of stresses on the surface of the deposited metal particles may facilitate the occurrence of a localized shear deformation that can cause microcracks on the surface of the newly produced metallic coating.

3.4. Microstructure Titanium Coatings

The SEM microstructure of the cross-section of the titanium coatings onto Al7075 substrate (for the of SoD of 20 mm, 70 mm, and 100 mm) before and after exposure in 1.2 M Cl^- solution (pH 1.5) are shown in Figure 9. The exposure time was 4 h.

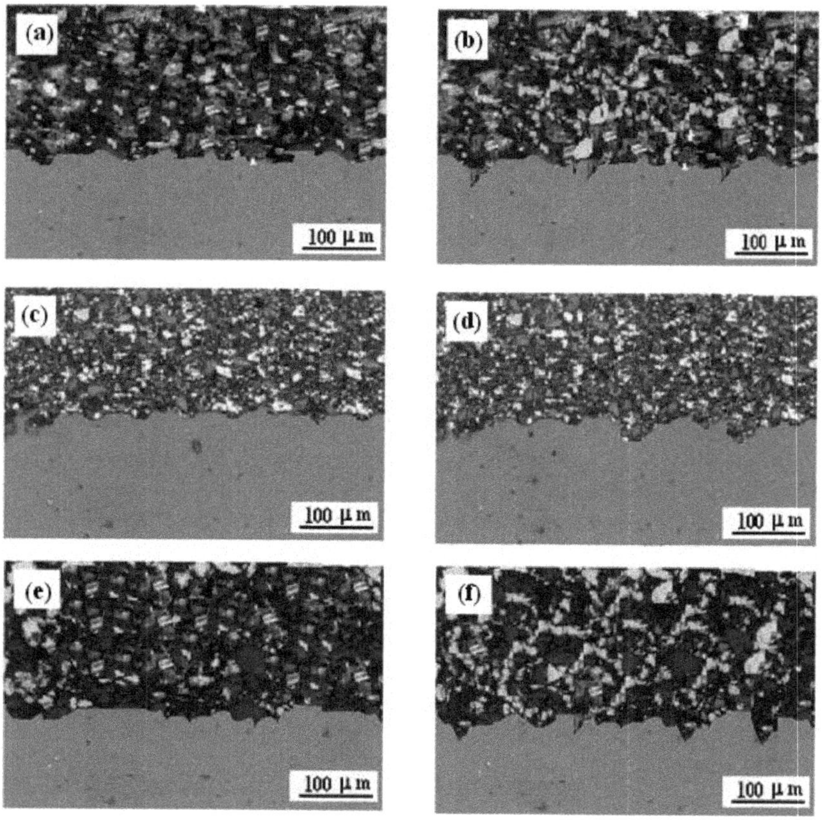

Figure 9. SEM of cross-section of titanium deposited onto Al7075 substrate at the SoD of: (**a**) 20 mm, (**c**) 70 mm, and (**e**) 100 mm, and after exposure at 1.2 M Cl^- (pH 1.5) at SoD of: (**b**) 20 mm, (**d**) 70 mm, and (**f**) 100 mm. Exposure time was 4 h.

Significant changes in the microstructure of the Al7075/Ti coatings, together with increasing of the SoD in the range of 20 mm to 100 mm, are visible. Spraying at short distances (20 mm) produces porous of Ti coatings which adhere well to the Al7075 alloy substrate (Figure 9a). The reason for such increased porosity of the coatings in these cases may be the lower speed of larger particles of the titanium powder. After leaving the nozzle, the velocity of the Ti powder grains increases depending on their size, reaching the highest value at a distance of 70 mm or 100 mm [25]. A close SoD is sufficient for large particles to form good adhesion with the substrate, with no pores visible at the interface. Extending the distance of the nozzle from the substrate leads to a noticeable reduction in the porosity, which reaches the minimum value at 70 mm, which is clearly visible in Figure 9c. However, this coating is characterized by a small number of very fine pores that appear evenly over the entire cross-sectional area.

Further increasing the distance of the nozzle from the substrate material, i.e., 100 mm, resulted in a significant increase in the porosity of the Ti coating onto Al7075 substrate (Figure 9e). The large distance of the nozzle from the substrate resulted in a decrease in the velocity of all the grains of the Ti powder, which resulted in an increase in the porosity of the Al7075/Ti coating. The most significant decrease in velocity concerned fine Ti powder grains, which deformed slightly, contributing to a significant increase in the porosity of the coating [31]. The kinetic energy of the largest Ti grains was so high that after impact with the previously applied layer of titanium, the coating deformation process took place. Such changes in the porosity of the coatings sprayed with cold gas result from the course of the coating forming process, where the last layer is not subjected to penning as intensively as the layers previously deposited [32,33].

The cross-section of the titanium coatings onto Al7075 substrate (for the SoD of 20 mm, 70 mm, and 100 mm) after exposure (4 h) in chloride solution are shown in Figure 9b–f. As a result of the corrosive action of the environment, the greatest destruction of the Ti layer was observed in the case of Al7075/Ti-20 and Al7075/Ti-100 samples (Table 1). It is clearly visible that the inhomogeneous, porous structure of the Ti coatings on the aluminum alloy substrate (Figure 9b,f) were degraded as a result of the reaction of the coating with the corrosive environment. The numerous pores on the titanium surface cause the electrolyte to penetrate deep into the Ti layer, which causes intense electrochemical corrosion of the aluminum substrate (reactions (6)–(8)). However, in the case of the SoD of 70 mm, a fairly smooth, compact structure of the Al7075/Ti coating was obtained. Therefore, the titanium surface was slightly damaged (Figure 9d), still protecting the Al7075 substrate from the corrosive action of the acid chloride solution.

4. Conclusions

The work carried out presents experimental studies of the titanium deposition onto Al7075 substrate by cold spraying at different distances between the nozzle and the substrate. The microstructure and anti-corrosion properties of the Ti onto Al7075 alloy in an acid chloride solution were investigated. The conducted research allowed for the formulation of the following conclusions:

1. The different standoff distance of the nozzle from the specimen surface (i.e., 20 mm, 70 mm, and 100 mm) in the cold spray process has a significant influence on the properties of the Al7075/Ti coating.
2. Titanium coatings adhere well onto Al7075 alloy, and the most homogeneous and smooth surface of the Ti coating was obtained when the nozzle distance from the sample surface was 70 mm.
3. The microhardness (HV0.3) of the deposit depend significantly on the nozzle distance. The highest level of HV0.3 value was achieved for deposits obtained with the SoD of 70 mm.
4. There were no phase changes in the phase composition of the titanium deposits due to the increased of SoD.
5. Corrosion test (electrochemical method) of the titanium coatings onto Al7075 substrate were carried out in acidic chloride solutions.
6. The mechanism of electrochemical corrosion of titanium coatings is a multi-stage process, and the main product of the corrosion process was $(TiO_2)_{ads}$. The oxide layer did not protect the materials against the penetration of the aggressive solution.
7. The polarization resistance (R_p) of the Ti coatings was the highest, while the corrosion rate (υ_{corr}) was the lowest, for the SoD of 70 mm. Thus, in this case, the exchange

of mass and electrical charge between the electrode and the electrolyte solution is significantly impeded.

8. The titanium surface on the Al7075 substrate was slightly damaged when exposed to an acid chloride solution, while still protecting the aluminum substrate from the corrosive effects of the environment.

Author Contributions: Conceptualization, M.S.; formal analysis, M.S. and K.S.-S.; investigation, K.S.-S.; methodology, M.S.; writing-original draft, M.S. and K.S.-S.; writing-review and editing, M.S. All authors have read and agreed to the published version of the manuscript.

Funding: The work reported herein has been undertaken as part of the project funded by the Jan Kochanowski University in Kielce, ID: SUPB. RN. 21.186.

Institutional Review Board Statement: Not applicable.

Informed Consent Statement: Not applicable.

Data Availability Statement: Data sharing is not applicable.

Conflicts of Interest: The authors declare no conflict of interest.

References

1. Assadi, H.; Gärtner, F.; Stoltenhoff, T.; Kreye, H. Bonding mechanism in cold gas spraying. *Acta Mater.* **2003**, *51*, 4379–4394. [CrossRef]
2. Li, W.-Y.; Gao, W. Some aspects on 3D numerical modeling of high velocity impact of particles in cold spraying by explicit finite element analysis. *Appl. Surf. Sci.* **2009**, *255*, 7878–7892. [CrossRef]
3. Yin, S.; Cavaliere, P.; Aldwell, B.; Jenkins, R.; Liao, H.; Li, W.; Lupoi, R. Cold spray additive manufacturing and repair: Fundamentals and applications. *Addit. Manuf.* **2018**, *21*, 628–650. [CrossRef]
4. Klinkov, S.V.; Fedorovich, V.F.; Rein, M. Cold spray deposition: Significance of particle impact phenomena. *Aerosp. Sci. Technol.* **2005**, *10*, 582–591. [CrossRef]
5. Raletz, F.; Vardelle, M.; Ezo'o, G. Critical particle velocity under cold spray conditions. *Surf. Coat. Technol.* **2006**, *201*, 1942–1947. [CrossRef]
6. Pattison, J.; Khan, A.; O'Neill, W.; Celetto, S. Standoff distance and bow shock phenomena in the cold spray process. *Surf. Coat. Technol.* **2008**, *202*, 1443–1454. [CrossRef]
7. Gnanasekaran, B.; Liu, G.R.; Fu, Y.; Wang, G.; Niu, W.; Lin, T. A Smoothed Particle Hydrodynamics (SPH) procedure for simulating cold spray process – A study using particles. *Surf. Coat. Technol.* **2019**, *377*, 24812. [CrossRef]
8. Hussain, T.; McCartney, D.G.; Shipway, P.H.; Zhang, D. Bonding mechanisms in cold spraying: The contributions of Metallurgical and Mechanical Components. *J. Thern. Spray Technol.* **2009**, *48*, 364–379. [CrossRef]
9. Schmidt, T.; Gartner, E.; Assadi, H.; Kreye, H. Development of a generalized parameter window for cold spray deposition. *Acta Mater.* **2006**, *54*, 729–742. [CrossRef]
10. Li, W.Y.; Liao, H.L.; Li, C.J.; Bang, H.-S.; Coddet, C. Numerical simulation of deformation behavior of Al particles impacting on Al substrate and effect of surface oxide films on interfacial bonding in cold spraying. *Appl. Surf. Sci.* **2007**, *253*, 5084–5091. [CrossRef]
11. Vidaller, M.V.; List, A.; Gaertner, F.; Klassen, T.; Dosta, S.; Guilemany, J.M. Single impact bonding of cold sprayed Ti-6Al-4V powders on different substrates. *J. Therm. Spray Technol.* **2015**, *24*, 644–658. [CrossRef]
12. Bhattiprolu, V.S.; Johnson, K.W.; Ozdemir, O.C.; Crawford, G.A. Influence of feedstock powder and cold spray processing parameters on microstructure and mechanical properties of Ti-6Al-4V cold spray depositions. *Surf. Coat. Technol.* **2018**, *335*, 1–12. [CrossRef]
13. Bhattiprolu, V.S.; Johnson, K.W.; Crawford, G.A. Influence of powder microstructure on the microstructural evolution of as-sprayed and heat treated cold-sprayed Ti-6Al-4V coatings. *J. Therm. Spray Technol.* **2018**, *28*, 174–188. [CrossRef]
14. Hajipour, H.; Abdollah-zadeh, A.; Assadi, H.; Taheri-Nassaj, E.; Jahed, H. Effect of feedstock powder morphology on cold-sprayed titanium dioxide coatings. *J. Therm. Spray Technol.* **2018**, *27*, 1542–1550. [CrossRef]
15. Munagala, V.N.V.; Akinyi, V.; Vo, P.; Chromik, R.R. Influence of powder morphology and microstructure on the cold spray and mechanical properties of Ti-6Al-4V coatings. *J. Therm. Spray Technol.* **2018**, *27*, 827–842. [CrossRef]
16. Munagala, V.N.V.; Bessette, S.; Gauvin, R.; Chromik, R.R. Sliding wear of cold sprayed Ti-6Al-4V coatings: Effect of porosity and normal load. *Wear* **2020**, *450–451*, 203268. [CrossRef]
17. Sirvent, P.; Garrido, M.Á.; Sharp, J.; Rainforth, W.M.; Poza, P. Improving the oscillating wear response of cold sprayed Ti-6Al-4V coatings through a heat treatment. *Surf. Coat. Technol.* **2020**, *399*, 126128. [CrossRef]
18. Wu, K.; Sun, W.; Tan, A.W.-Y.; Marinescu, I.; Liu, E.; Zhou, W. An investigation into microstructure, tribological and mechanical properties of cold sprayed Inconel 625 coatings. *Surf. Coat. Technol.* **2021**, *424*, 127660. [CrossRef]
19. Wu, K.; Chee, S.W.; Sun, W.; Tan, A.W.-Y.; Tan, S.C.; Liu, E.; Zhou, W. Inconel 713C coating by cold spray for surface enhancement of Inconel 718. *Metals* **2021**, *11*, 2048. [CrossRef]

20. Samareh, B.; Stier, O.; Lüthen, V.; Dolatabadi, A. Assessment of cfd modeling via flow visualization in cold spray process. *J. Therm. Spray Technol.* **2009**, *18*, 934–943. [CrossRef]
21. Meyer, M.C.; Yin, S.; Lupoi, R. Particle in-flight velocity and dispersion measurements at increasing particle feed rates in cold spray. *J. Therm. Spray Technol.* **2016**, *26*, 60–70. [CrossRef]
22. Meyer, M.C.; Yin, S.; McDonnell, K.A.; Stier, O.; Lupoi, R. Feed rate effect on particulate acceleration in cold spray under low stagnation pressure conditions. *Surf. Coat. Technol.* **2016**, *304*, 237–245. [CrossRef]
23. Morgan, R.; Fox, P.; Pattison, J.; Sutcliffe, C.; O'Neill, W. Analysis of cold gas dynamically sprayed aluminium deposits. *Mater. Lett.* **2004**, *58*, 1317–1320. [CrossRef]
24. Grujicic, M.; Zhao, C.L.; DeRosset, W.S.; Helfritch, D. Adiabatic shear instability based mechanism for particles/substrate bonding in the cold-gas dynamic-spray process. *Mater. Des.* **2004**, *25*, 681–688. [CrossRef]
25. Żórawski, W.; Molak, R.; Mądry, J.; Sienicki, J.; Góral, A.; Makrenek, M.; Scendo, M.; Dobosz, R. Experimental and numerical investigations of titanium deposition for cold spray additive manufacturing as a function of standoff distance. *Materials* **2021**, *14*, 5492. [CrossRef] [PubMed]
26. Scendo, M.; Zorawski, W.; Staszewska-Samson, K.; Goral, A. Influence of laser treatment on the corrosion resistance of Cr_3C_2-25(Ni20Cr) cermet coating. *Materials* **2021**, *14*, 4078. [CrossRef] [PubMed]
27. Scendo, M.; Radek, N.; Trela, J. Influence of laser treatment on the corrosive resistance of WC-Cu coating produced by electrospark deposition. *Int. J. Electrochem. Sci.* **2013**, *8*, 9264–9277.
28. Scendo, M.; Trela, J.; Radek, N. Influence of laser power on the corrosive resistance of WC-Cu coating. *Surf. Coat. Technol.* **2014**, *259*, 401–407. [CrossRef]
29. Scendo, M.; Zorawski, W.; Staszewska, K.; Makrenek, M.; Goral, A. Influence of surface pretreatment on the corrosion resistance of cold spray nickel coatings in acidic chloride solution. *J. Mater. Eng. Perform.* **2018**, *27*, 1725–1737. [CrossRef]
30. Adjdelsztajn, L.; Jodoin, B.; Lavernia, E. Cold gas dynamic spraying of a high temperature Al alloy. *Surf. Coat. Technol.* **2006**, *201*, 2109–2116. [CrossRef]
31. Li, W.-Y.; Zhang, C.; Wang, H.-T.; Guo, X.-P.; Liao, H.L.; Li, C.-J.; Coddet, C. Significant influences of metal reactivity and oxide films at particle surfaces on coating microstructure in cold spraying. *Appl. Surf. Sci.* **2007**, *253*, 3557–3562. [CrossRef]
32. Li, W.-Y.; Zhang, C.; Guo, X.P.; Zhang, G.; Liao, H.L.; Li, C.-J.; Coddet, C. Effect of standoff distance on coating deposition characteristics in cold spraying. *Mater. Des.* **2008**, *29*, 297–304. [CrossRef]
33. Scendo, M.; Zorawski, W.; Goral, A. Influence of Nickel Powders on Corrosion Resistance of Cold Sprayed Coatings on Al7075 Substrate. *Metals* **2019**, *9*, 890. [CrossRef]

coatings

MDPI

Article

Quantitative Characterization of Interfacial Defects in Thermal Barrier Coatings by Long Pulse Thermography

Jinfeng Wei [1,2,†], Guangnan Xu [1,†], Guolin Liu [1], Jinwei Guo [1,*], Wang Zhu [1,*] and Zengsheng Ma [1]

1 School of Materials Science and Engineering, Xiangtan University, Xiangtan 411105, China
2 Hunan Aviation Powerplant Research Institute, Aero Engine Corporation of China, Zhuzhou 412000, China
* Correspondence: jwguo@xtu.edu.cn (J.G.); wzhu@xtu.edu.cn (W.Z.)
† These authors contributed equally to this work.

Abstract: The non-contact long pulse thermography method is commonly used to detect the defects in thermal barrier coatings (TBCs). The profile of interfacial defect in TBCs can be monitored by infrared camera under the irradiation of the excitation source. Unfortunately, the defect profile is always blurry due to heat diffusion between the defect area and the intact area. It is difficult to quantify the size of defect size in TBCs. In this work, combined with derived one-dimensional heat conduction analytical model, a non-contact long pulse thermography (LPT) method is applied to quantitatively investigate the interface defects in TBCs. Principal component analysis (PCA) and background subtraction method are used to improve the contrast of the defect profile in collected thermal images. By fitting the results between the profile of the interface defect in thermal images and the predicted shape of the model, the interface defect size can be determined. Furthermore, a simple extension of proposed method for interfacial defects with irregular shape is presented. The predicted errors for round defect with diameters of 3 mm, 5 mm and 7 mm are roughly distributed in the range of 3%~6%, which are not affected by the defect diameter.

Keywords: thermal barrier coatings; long pulse thermography; interfacial defects; fitting background subtraction; quantitative characterization

1. Introduction

Thermal barrier coatings (TBCs) are widely used to improve the temperature-bearing capacity of high temperature components in aircraft engines such as high pressure turbine blades and combustors [1,2]. A typical system of TBCs is composed of a superalloy substrate, a ceramic topcoat with low thermal conductivity, and a metallic bond coat for offering adhesion between the topcoat and the substrate [3–6]. In addition, thermally grown oxide (TGO) is formed in the interface between the ceramic topcoat (TC) and bond coat (BC) due to oxidation at high temperature [7]. TBCs can effectively provide reductions in the surface temperature of the superalloy substrate (100–300 °C) [8]. Due to the complex multi-layer structures of TBCs and the harsh service environment, the coating is prone to interfacial defects such as debonding and buckling. Further propagation of these defects will lead to spalling failure of TBCs, which seriously threatens the safe application of TBCs on turbine blades. Many previous studies show that the spallation of TBCs starts as a separation between TC and BC. Zhen et al. [9] found that delamination cracking appears in the interior of the TGO layer and the interface area between BC and TGO layers after thermal cycling. Wei et al. [10] showed that the surface vertical crack expands to the interface of BC/TGO layers and merges rapidly with the horizontal crack, with the consideration of TGO growth and ceramic sintering. As the delamination between TC and BC becomes large enough, buckling appears [11,12], and enough buckling will cause the spalling of TC. However, some interfacial defects are buried in the bottom of the ceramic coating, which are hard to directly observe with the naked eye. In order to avoid sudden accidents and loss due to the detachment of TBCs, it is very important to develop an effective way to predict and evaluate the state of defect in TBCs.

Over the years, several nondestructive techniques such as ultrasonic [13], impedance spectroscopy [14], X-ray computed tomography [15,16], and infrared thermography nondestructive testing [17,18] have been used to evaluate and identify the state of defect in TBCs. Impedance spectroscopy and ultrasonic testing are the contact detection methods, and they are not conducive to defect detection for TBCs in many in-service situations. X-ray computed tomography is able to analyze the 3D spatial microstructure in nondestructive mode, but the inspection equipment is relatively costly, and the top ceramic coating need to be removed from the superalloy substrate for alleviating the effect of X-ray absorption. The above mentioned nondestructive testing methods are not ideal for the quantitative characterization of internal defects in TBCs. Infrared thermography testing has become a hot defect evaluation method due to the advantages of non-contact, high efficiency, direct detection results, and quantitative characterization of defects [19].

Long pulsed thermography (LPT) is commonly used to detect internal defects [20]. The detection principle of infrared thermography is that the excitation source (such as flash bulbs, halogen lamps) heats the surface of the sample and the temperature change of the sample surface is monitored and recorded by an infrared camera. Due to the difference of the thermal properties of defective and intact region, the temperature in the defective region is different from that in the intact region. Then, the information of defects can be detected, such as debond, crack, uneven thickness of coating, delamination and so on. Newaz et al. [21] investigated the progressive damage assessment in TBCs using PT technique, and the results verified that coating layering and buckling happens only toward the end of life and has a very short span. Combining Terahertz-Time domain spectroscopy (THz-TDS) with the PT technique, the degree of degradation of the thickness of TBCs top coat of varying service lifespans were quantitatively analyzed [22]. Based on the thermography method, Kumar et al. [23] collected the thermographic images of TBCs samples exposed to thermal cycling. The experimental results are consistent with the FEM results, which verified the delamination form of defect. Shi et al. [24] developed one-way coupling algorithm (GEMSS) and carried out the investigation of the thermal image collected by infrared thermography to detect cracks in TBCs. The amount and position of crack with different lengths can be identified from collected thermal image. Zhu et al. [25] combined the acoustic emission method with infrared thermography to explore the degradation mechanism of TBCs corroded by CMAS under thermal shock. The increased abnormal temperature area with the number of thermal shock cycles can be observed from the thermal images. The abnormal area was recognized as the spalling location.

Recently, a few quantitative evaluation methods of internal defect size have been developed for TBCs. Zhuo et al. [26] developed a size estimation method of interior defects, which added a hypothetical heat flux supplied from surface temperature measured by step heating thermography. Guo et al. [27] studied the diameter of in-plane defect by LPT, and proposed a novel method FWHM-1st to quantify the diameter. The error of their measurement results of defect becomes larger with the decrease in the defect size, and the minimum error reached 12%. Due to the inevitable fact that heat diffusion blurs the defect boundary in thermal images, the accurate evaluation of the sizes of smaller defects still needs to be developed.

In order to further enhance the assessment ability of the size of interfacial defects in thermal images of TBCs, we proposed a quantification method by fitting the results between the profile of interface defect in thermal images and the predicted shape of the model in this paper. First, LPT was used to quantitative investigate the interfacial defects of TBCs. Then, PCA is applied to extract a reconstructed thermal image from the abundant thermal images collected by LPT. The background subtraction method was used to reduce the background influence in the collected thermal images. Furthermore, a one-dimensional heat conduction analytical model of TBCs with interface defects was derived. By fitting the results between the profile of interface defect in thermal images and the predicted shape of the model, the interface defect size can be determined. At last, a simple extension of current evaluation method of defect size was presented for interfacial defects with irregular shape.

2. Experiments

2.1. Sample

Two flat-plate test samples with the dimensions of 115 mm × 40 mm ×5 mm were prepared for the experiment. Each sample was composed of the stainless steel substrate, precast brass defects, and 7 wt.% yttria-stabilised zirconia (YSZ) coating. In order to produce interfacial defects in the samples, several cylindrical holes with different diameters and the same depth of 1.5 mm were drilled in the stainless steel substrate before spraying. Then, these holes were plugged with copper, which had the same dimensions as holes. The height of copper billet was slightly less than the depth of the cylindrical holes, therefore, a paper-thin air gap existed in the interface between the copper billet and stainless steel. Then, the surface of the stainless steel substrate was polished flat. Finally, the YSZ coating with a thickness of 200 μm was prepared on the surface of the substrate by air plasma spraying. The processing parameters for spraying the YSZ coating are given in Table 1.

Table 1. Spraying parameter for YSZ coat.

Voltage (V)	Current(A)	Primary Gas Ar (slpm)	Primary Gas H_2 (slpm)	Standoff Distance (mm)	Feed Rate (g/min)
75	600	40	15	110	20

Sample 1 is used to propose the image processing method and the quantitative method of the size of defect. Sample 2 is used to verify the quantitative method. Figure 1 shows the schematic diagram of the dimension of the sample 1. The center of defects was in the bisector of substrate. The samples with different sizes of defects were used to investigate the effect of defect diameters on temperature distribution law. The dimensions of the in-plane defects are given in Table 2.

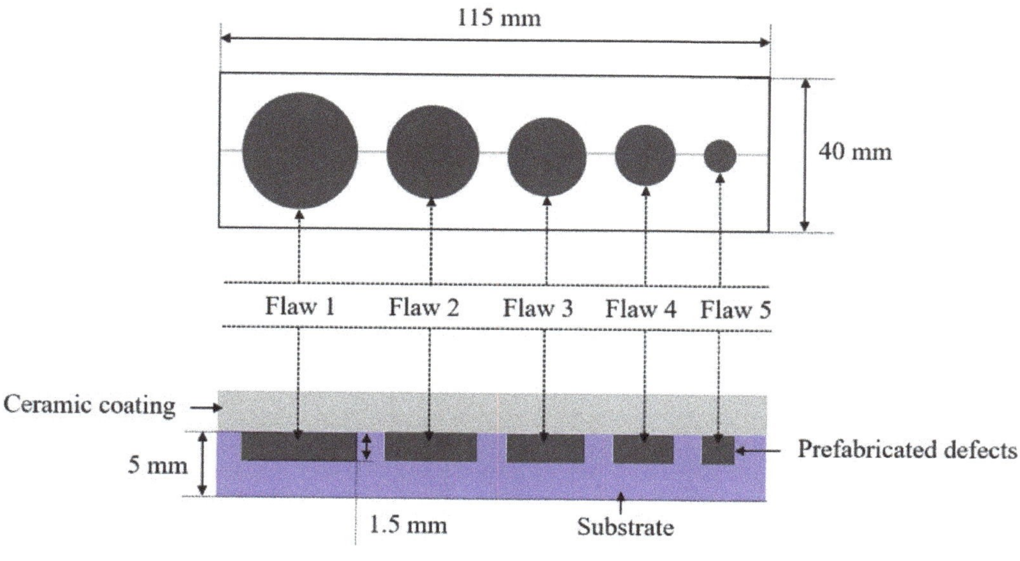

Figure 1. The schematic diagram of the dimension of the sample 1.

Table 2. Thickness of ceramic coating and the dimension of the interfacial defects.

Sample Number	Diameter of Defect *D* (mm)
1	Flaw 1: 10; Flaw 2: 8; Flaw 3: 6; Flaw 4: 4; Flaw 5: 2
2	Flaw 1: 7; Flaw 2: 5; Flaw 3: 3

2.2. Experimental Setup and Data Collection

A flow diagram of defect detection by LPT is shown in Figure 2. Two halogen lamps with a high power of 2 kW were selected as the excitation sources. In order to reduce the effect of uneven heating, the two lamps were placed symmetrically on both sides of the test sample, and the space between them was about 1 m. An infrared camera was put in the middle of the two lamps to capture the temperature evolution of samples surface. The objective table was used to place the tested sample, and located at the right ahead of the infrared camera. The spectral range, thermal sensitivity at room temperature and resolution of the infrared camera were 3.8~4.05 μm, 25 mK, and 320 × 240 pixels, respectively. In this study, the sampling frequency was 30 Hz, the heating time was 6 s, and the distance between the sample and the infrared camera was 30 cm.

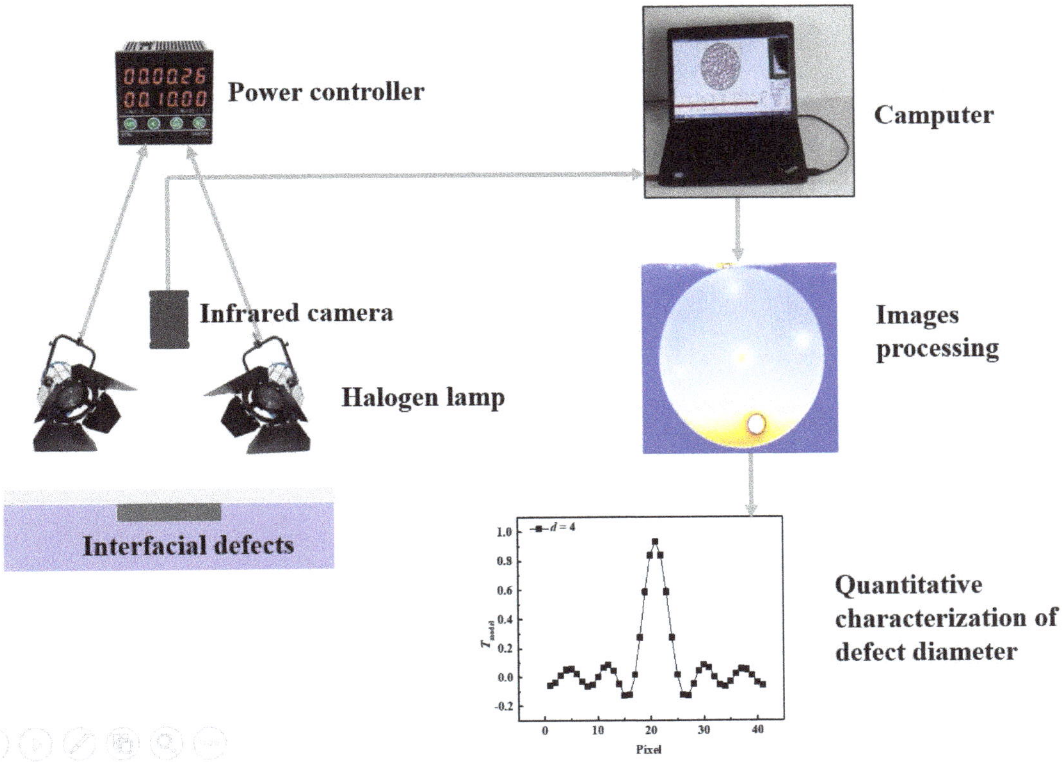

Figure 2. A flow diagram of defect detection by LPT.

Figure 3 shows an infrared image of the sample surface captured by the infrared camera. The temperature of the defective area was higher than that of the intact area. The 90 frames thermal images in the first 3 s period after removing the thermal excitation source were used to analyze the characteristic information of temperature. In order to reduce the effect of environmental noise on the defect identification, the first frame of the images was used as the background image, and the reconstructed 90 frames thermal sequence images were obtained by subtracting background image from the images, including noise interference signal. To facilitate the following processing, the ratio ($\overline{T}$) of temperature data T extracted from the subtracting background thermal sequence image to T_d (T_d = 1 °C) was obtained.

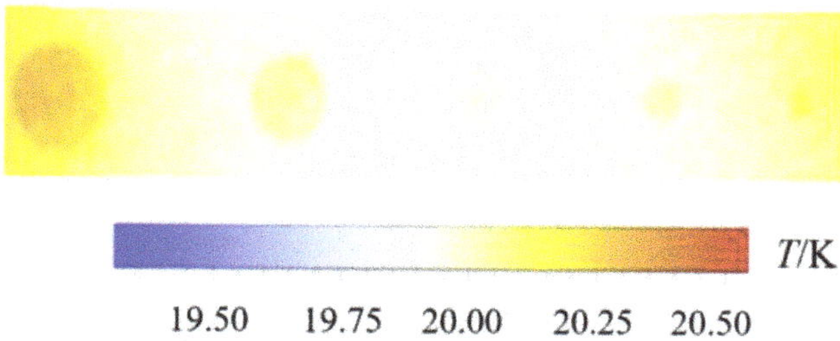

Figure 3. Thermal data captured by an infrared camera.

2.3. Principal Component Analysis (PCA)

PCA was used to process thermal image sequences with uneven heating noise and improve the defect detectability from multiple frames figures [28,29]. Based on the principle of data dimensionality reduction reconstruction, PCA uses the singular value decomposition (SVD) to extract the spatial and temporal information containing defect features from abundant thermal image sequences [28,30]. The SVD is written as follows:

$$[A] = [U][R][V]^T \tag{1}$$

where $[A]$ is a matrix, whose column vector is composed by each frame image compressed into a single column. $[U]$ and $[V]$ are the left and right singular matrixes of the matrix $[A]$. $[R]$ is a 90 × 90 diagonal matrix with the singular values of matrix $[A]$ in the diagonal and the elements of matrix $[R]$ are equal to zero on off-diagonal lines. Considering that data matrix $[A]$ is arranged time variations occur column-wise and spatial variations occur row-wise, the columns of matrix $[U]$ are composed by the empirical orthogonal functions (EOF) that can describe the spatial variables of the thermal response data. Matrix $[V]$ can describe the characteristic time behavior, where the principal component (PC) vectors are arranged row-wise [31,32].

Corresponding to the three principal components of the 90 frames of thermal images processed in previous section, the reconstructed thermal sequence images by PCA algorithm are shown in Figure 4. Conveniently, we denoted the reconstructed image of the n principal component as PCn. The temperature distribution in PCn is denoted by T_{PCn}. Figure 4a shows that a reconstructed thermal sequence image of the first principal corresponds to the three principal components of the 90 frames thermal images processed in the previous section; the reconstructed thermal sequence images by PCA algorithm are shown in Figure 4. Conveniently, we denoted the reconstructed image of the n principal component as PCn. The temperature distribution in PCn is denoted by T_{PCn}. Figure 4a shows that a reconstructed thermal sequence image of the first principal component PC1, which contains most of the complete defect characteristic information of the 90 frames thermal sequence images. It is easy to identify the defect location. The reconstructed thermal sequence image of the second principal component PC2 is greatly affected by noise, and some defects are difficult to identify (see Figure 4b). In Figure 4c, the defect shape in the reconstructed thermal sequence image of the third principal component PC3 is severely distorted. Therefore, the reconstructed thermal sequence image of PC1 is selected for the following procedure. Line 1 marked by a dotted line in Figure 4a is used to the following quantitative analysis of defects.

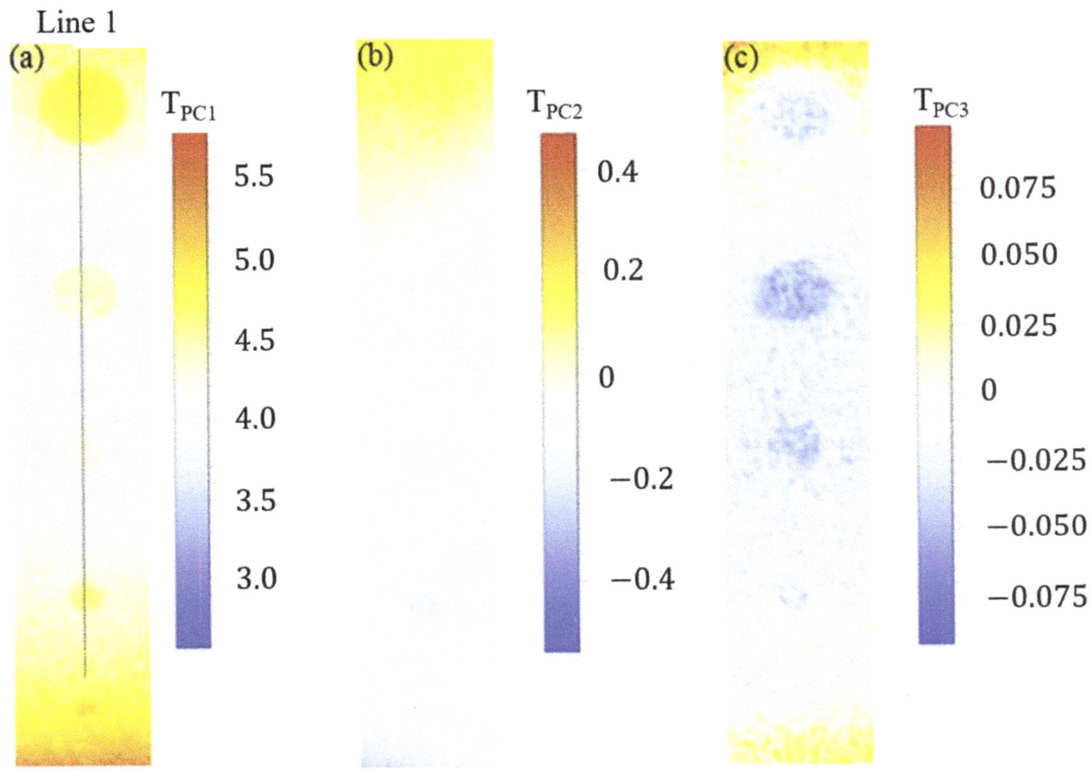

Figure 4. Reconstructed images by PCA algorithm of the sample 1: (**a**) PC1; (**b**) PC2; (**c**) PC3. Line 1 across the center of all flaws.

2.4. Subtracting Fitting Background

Two excitation lamps were placed symmetrically on both sides of the sample to reduce the effect of uneven heating. However, with the symmetrical distribution of the lamps, the effect of uneven heating was still existent. Figure 5 shows the temperature distribution along line 1 in Figure 4a. The temperature of the sample surface at the defect region is obviously higher than that at the intact region. Due to the uneven heating, the temperatures at both ends of the sample surface are higher, and the temperature in the central region is lower than that at both ends. To solve the adverse effect of uneven heating on defect detection, an image processing method of background subtraction is proposed in this section.

The process of subtracting fitting backgrounds consists of two steps: fitting background and background subtraction. Figure 6 shows the three-dimensional thermal images of flaw 1 in the sample 1. The three-dimensional size of defect can be identified as an abnormal rise region of temperature. The thermal data of the intact region is used to predict the thermal background in the defective region. A quadratic equation is proposed to fit the data of the thermal background:

$$T_{\text{fit}} = a + bx/l_{\text{p}} + c(x/l_{\text{p}})^2 \tag{2}$$

where T_{fit} is the thermal background fitted by the thermal data of the intact region in Figure 6a. a, b, and c are the fitting parameters. x denotes the pixel points and l_{p} is one pixel. Next, we subtract the raw PCA image sequence T_{PC1} with the thermal background T_{fit}:

$$T_{Pf} = T_{PC1} - T_{fit} \quad (3)$$

where T_{Pf} is the thermal data in pixel points after subtracting the fitting background.

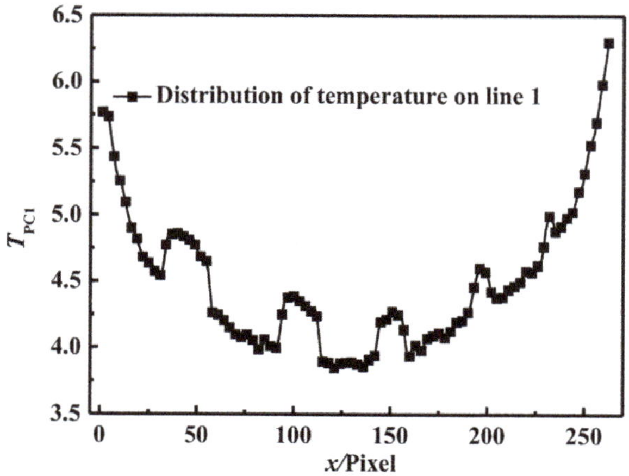

Figure 5. The temperature distribution along line 1 in Figure 4a.

Figure 6a shows a three-dimensional thermal image of PC1. The corresponding temperature values at two ends of the sample surface are different due to the uneven heating. To decrease this effect, a subtracting fitting background method is proposed. As shown in Figure 6b, a thermal image processed by the subtracting fitting background method can effectively reduce the local uneven heating noise. The processed images of this section are used to quantitative characterization analysis in the next section.

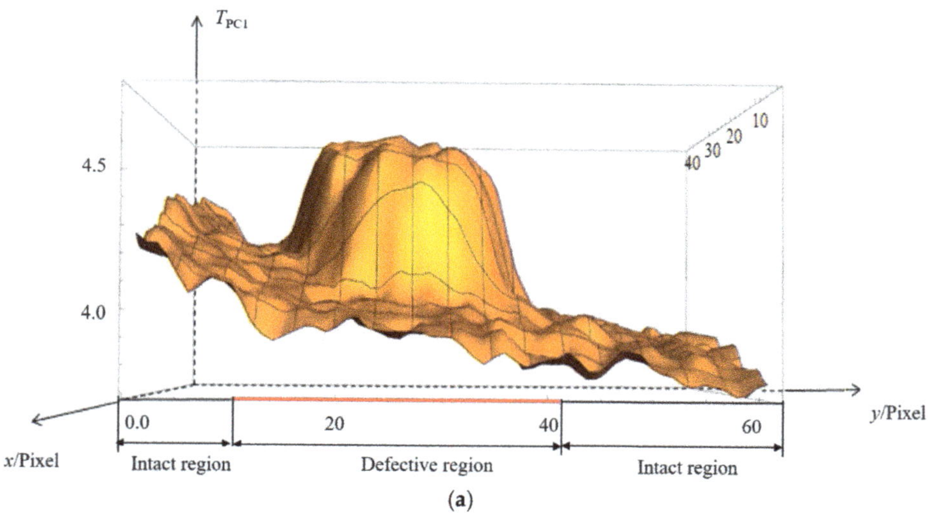

Figure 6. *Cont.*

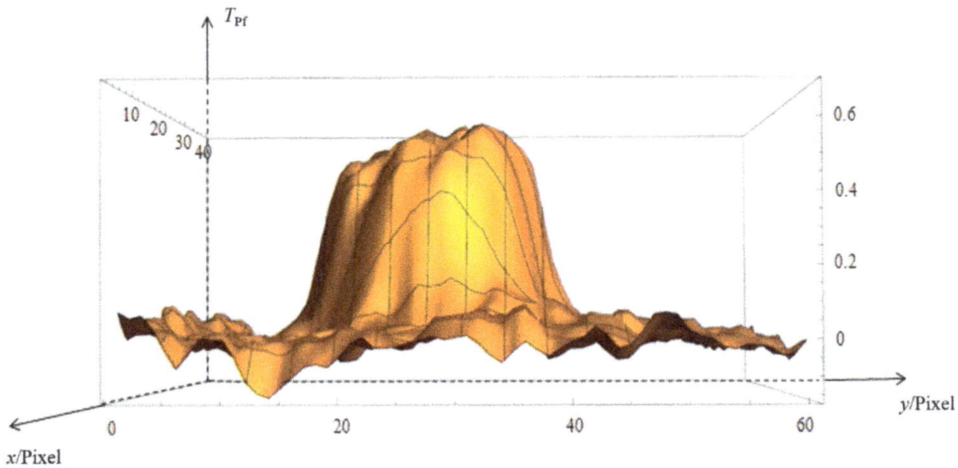

(**b**)

Figure 6. Three-dimensional thermal images of flaw 1 in the sample 1: (**a**) a raw thermal data extracted from PC1; (**b**) a thermal data extracted from PC1 after subtracting fitting background. Plane coordinates are represented by x-axis and y-axis. Vertical axis is used to record the value of temperature. Red line represents the coordinate range in y-axis, where the temperature has an abnormal rise.

3. Results and Discussion

3.1. Temperature Curve in the Defective Region after Subtracting the Fitting Background

To quantitatively analyze the lateral dimension of defects from the thermal images, the temperature distribution along line 1 was extracted from the thermal sequence image of PC1 after subtracting the fitting background. Figure 7 shows the temperature distribution in line 1 around defects with different diameters. The uneven background temperature can be effectively eliminated by algorithm processing of subtracting the fitting background. Due to the effect of the interface defect at the bottom of the ceramic coating on the distribution of the surface temperature of ceramic coating, the temperature in the defect area is higher than the intact region during the heating process by the heat source. After removing the source of motivation, the heat flows from the high temperature defect area to the low temperature intact area. Then, in collected infrared thermographs, the boundary of the high temperature area is fuzzy because the temperature changes from the heat flow. Also, the effect of interface defect on the transitive distance leads to the blurring of boundaries. Therefore, the sloping boundary of high-temperature area can be seen in Figure 7, and form the transition region between high and low temperature. When the lateral dimension of a defect is larger than 6 mm, the temperature distribution of the defective region presents a platform zone with undulating waveform. Note that the width of platform zone is a little less than the width of interfacial defect due to the transition region between high and low temperature. The width and height of the platform zone gradually decrease by decreasing the lateral dimension of a defect. When the lateral dimension of a defect is less than 6 mm, the transition regions distributed in the both sides of the high temperature area infinitely approach. Then, the platform zone disappears and reduces to an arch. The width and height of the arch also gradually decrease with decreasing the lateral dimension of a defect.

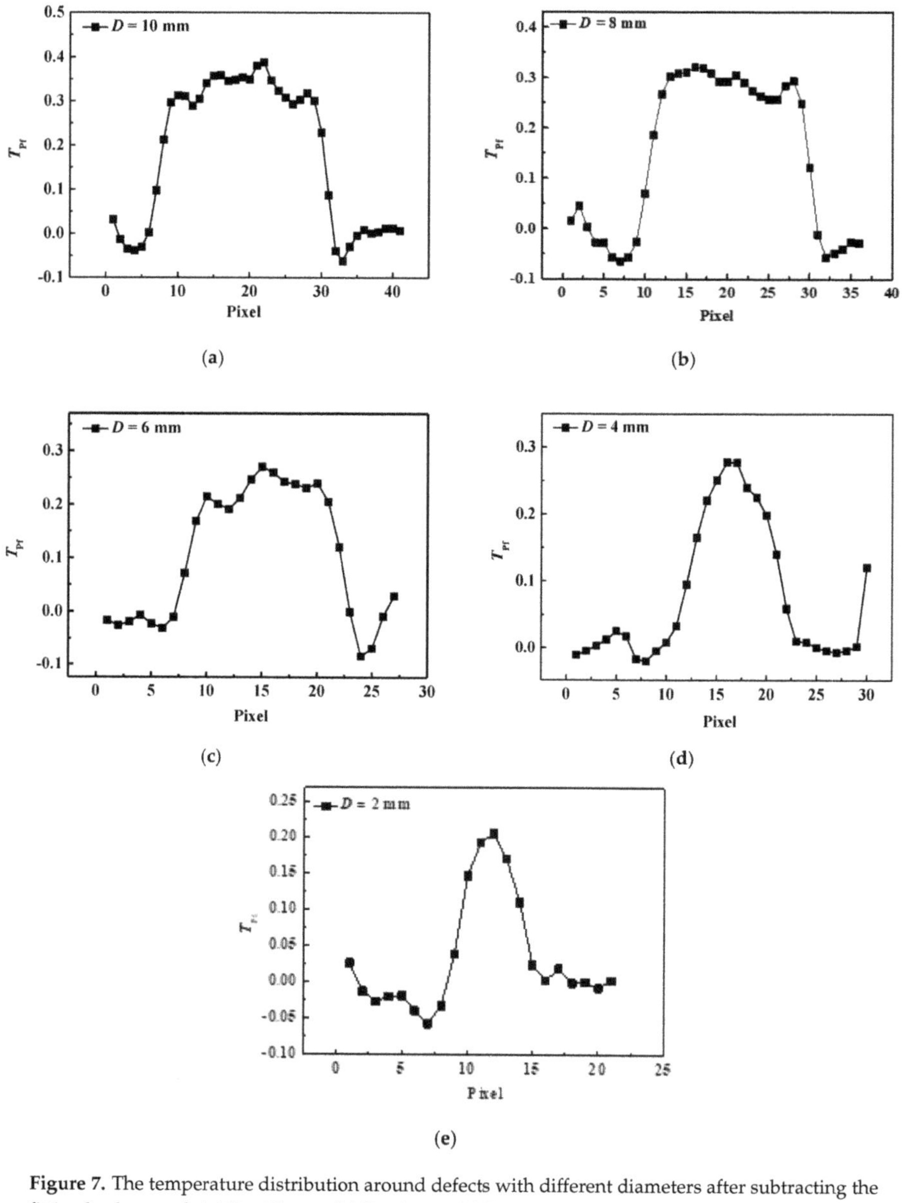

Figure 7. The temperature distribution around defects with different diameters after subtracting the fitting background: (**a**) D = 10 mm; (**b**) D = 8 mm; (**c**) D = 6 mm; (**d**) D = 4 mm and (**e**) D = 2 mm.

3.2. Heat Transfer Model of Interface Defect

The surfaces of defective and intact regions radiate heat outward during the sample heating by an excitation source. At the same time, the temperature difference between the defective region and the intact region leads the heat to flow transversely between the two regions (see Figure 8). Heat conduction can be represented by two processes, which are thermal diffusion in the vertical direction and transverse direction. According to the experimental results, the surface temperature of sample at the defective region is higher than that at the intact region during heating by an excitation source and cooling. Therefore, the temperature difference in the vertical direction is far less than that in the transverse direction. Coating material at the defective region can be seen as a heat source when the excitation source is turned off. The temperature difference at the moment just turning off the excitation source is the most obvious. After a period of cooling, the temperature difference disappears gradually. In this work, we ignore the effect of heat diffusion in the vertical direction, and assume an ideal situation that the temperature on the surface of ceramic coating has a distribution as shown in Figure 9 at the moment just turning off the excitation source. The superscripts defect and intact represent variable in the defective and intact region.

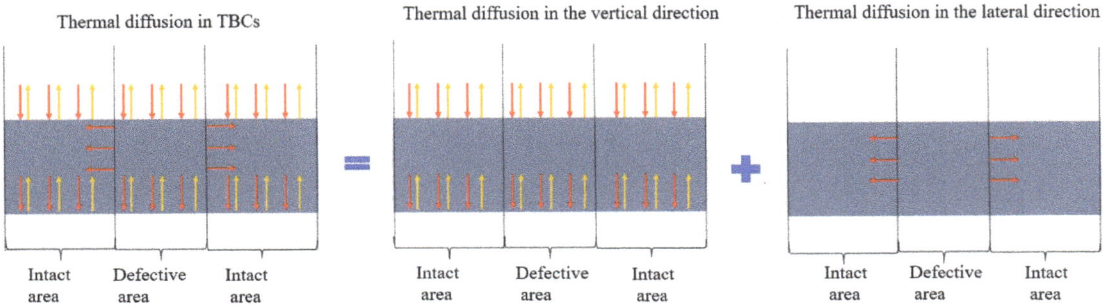

Figure 8. Schematic diagram of thermal diffusion in TBCs with interfacial defect.

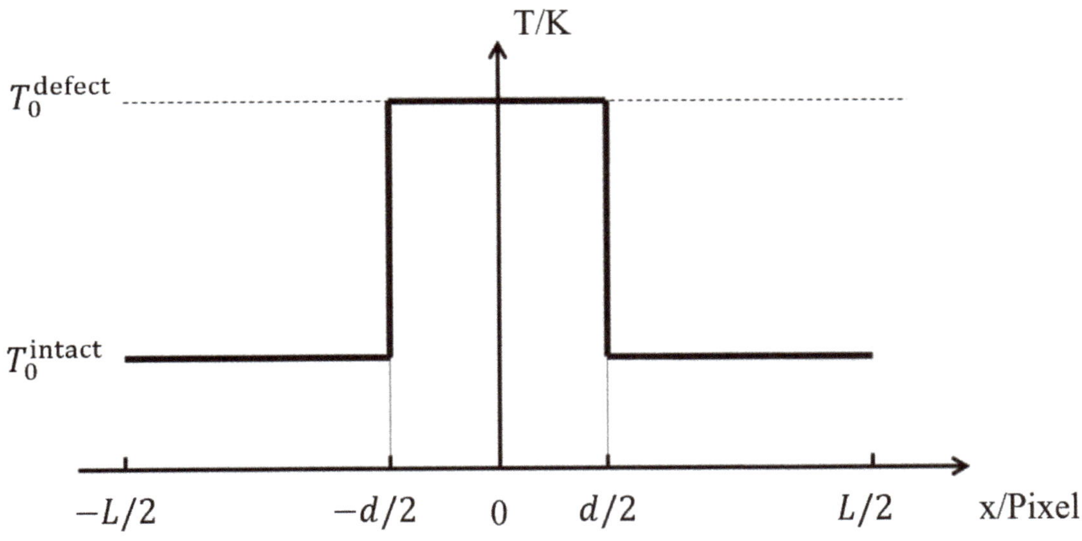

Figure 9. Temperature distribution on the surface of ceramic coating at the moment just turning off the excitation source.

A schematic diagram of heat conduction in the ceramic coating with interface defect during cooling process is shown in Figure 10. The differential equation of heat conduction on the surface of coating at the x axis direction is written as follows

$$\rho c \frac{\partial T_{\text{model}}(x, -e_1, t)}{\partial t} = \frac{\partial}{\partial x}\left(k_T \frac{\partial T_{\text{model}}(x, -e_1, t)}{\partial x}\right) \quad (4)$$

where ρ, c and k_T are the density, thermal capacity and the thermal diffusion coefficient of ceramic coating, respectively.

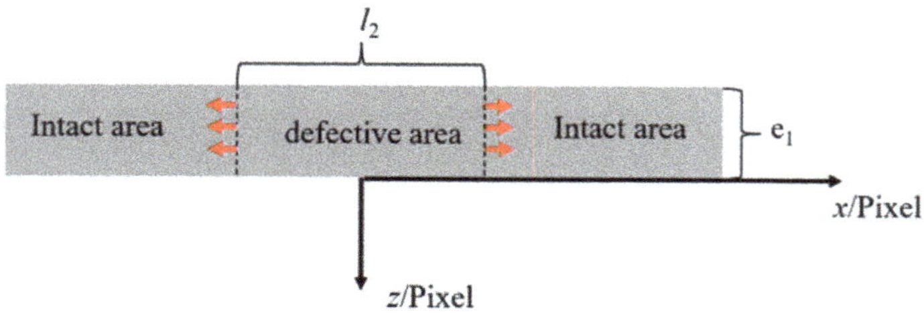

Figure 10. A schematic diagram of heat conduction in TBCs with defective area during cooling process. Red arrow denotes the direction of heat flow diffusion.

The initial condition and boundary condition are written as follows

$$T_{\text{model}}(x, -e_1, 0) = \phi(x) \tag{5}$$

$$T_{\text{model}}(-L/2, -e_1, t) = T_{\text{model}}(L/2, -e_1, t) = T_0^{\text{intact}} \tag{6}$$

where $\phi(x)$ is the temperature distribution on the surface of coating during cooling process and it presents a form of step function (see Figure 9). L and d are the length of the sample and the defect size in the model, respectively.

Then, an analytical expression of surface temperature can be derived as

$$\begin{aligned} T_{\text{model}}(x, -e_1, t) &= \sum_{n=1}^{k} A_n \exp\left(-a^2(nw)^2 t\right) \sin(nwx + \varphi) \\ A_n &= \frac{2}{L} \int_{-L/2}^{L/2} \phi(x) \sin(nwx + \phi) \mathrm{d}x \ \ n = 1,\ 2,\ 3,\ \ldots,\ k \end{aligned} \tag{7}$$

where a and φ are the constants. w is the fundamental frequency. Equation (7) is composed of cumulative sum and has a form of Fourier Series of $\phi(x)$.

After subtracting the fitting background, the temperature at the intact region T_0^{intact} is equal to zero. The temperature difference at the moment just turning off the excitation source is the most obvious. Then, we assume that the cooling time t is infinitely near zero. Then Equation (7) can be simplified as the discrete form of $\phi(x)$ and written as

$$\begin{aligned} T_{\text{model}} &= T_0^{\text{defect}} + \sum_{n=1}^{m} T_0^{\text{defect}} a_n \cos(2nx\pi/L) + T_0^{\text{defect}} b_n \sin(2nx\pi/L) \text{ for } n = 1,\ 2,\ 3, \ldots, m \\ a_n &= \frac{1}{L} \int_{-L/2}^{L/2} T \cos(2nx\pi/L) \mathrm{d}x \\ b_n &= \frac{1}{L} \int_{-L/2}^{L/2} T \sin(2nx\pi/L) \mathrm{d}x \end{aligned} \tag{8}$$

Keep the value m as five, the variation of temperature function curve with different values of d is shown in Figure 11. It is worth noting that the changing trend of the temperature function curves with the decrease of d is the same as that shown in Figure 7. When the value of d is larger than 12, the distribution of T_{model} at the area between $-d/2$ and $d/2$ presents a platform zone with undulating waveform. The width and height of the platform zone gradually decrease with decreasing d. When the value of d is less than 12, the platform zone reduces to an arch. After that, the width and height of the arch also gradually decrease with decreasing d.

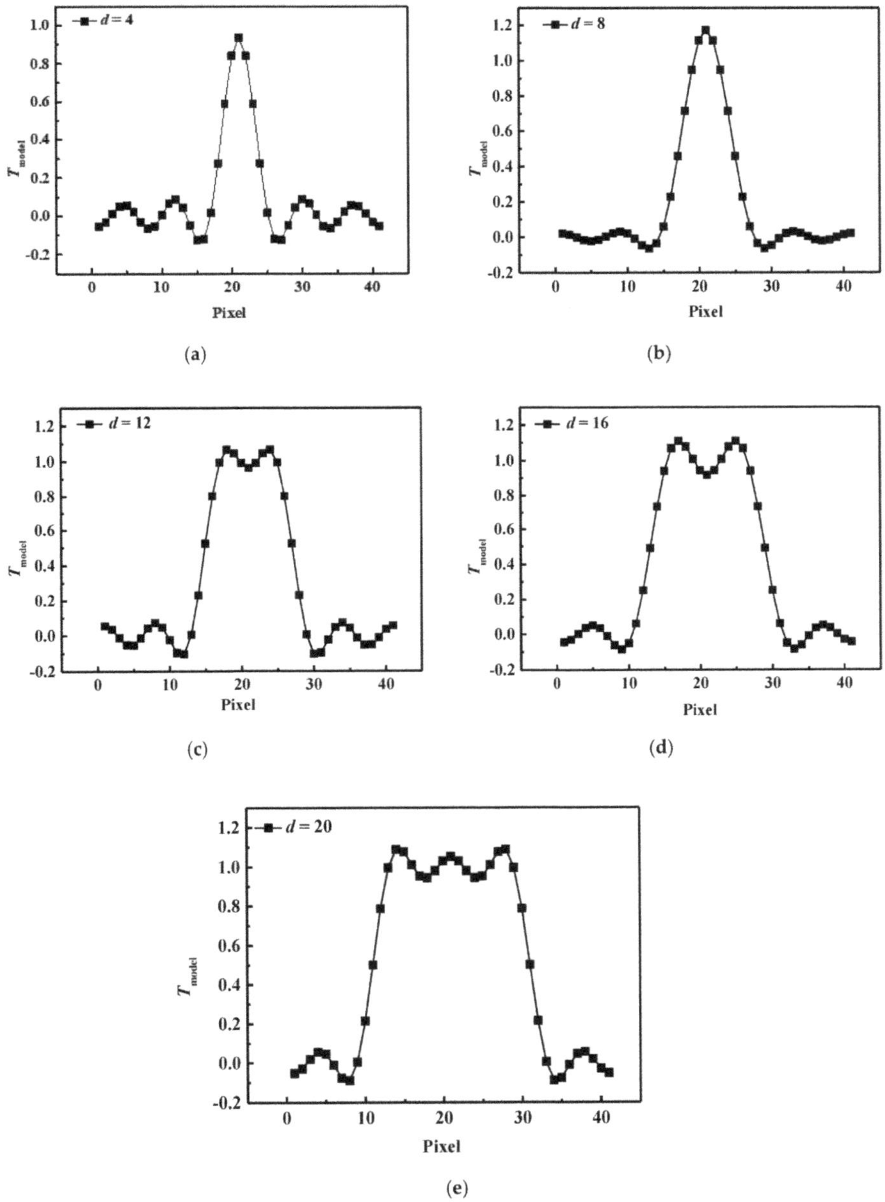

Figure 11. The discrete form of the Fourier series expansion of Equation (8) with different value of d: (**a**) d = 4 pixel; (**b**) d = 8 pixel; (**c**) d = 12 pixel; (**d**) d = 16 pixel and (**e**) d = 20 pixel.

3.3. Quantitative Analysis of Defect Size

Here we obtain the value of d corresponding to the defect size D of the flaw in sample 1 by editing the arithmetic to fit the Equation (8), with temperature curve shown in Figure 7. It should be noted that the change of curve shape from platform zone to arch with the decrease of d is remarkable, as shown in Figures 7 and 11. Therefore, only the change of the curve shape with different d is used to identify the corresponding D in this work. To avoid the effect of height difference between function curve of the model and temperature curve of the experiment on quantitative analysis of defect size, we make the parameter T_0^{defect} automatically achieve the best match when the value of *d* is fixed.

Figure 12 gives the temperature curves of the experiment and the corresponding function curves with different *d*. R-square is used to represent the degree of fitting between curves of the model and the experiment. The right of Figure 12a shows the values of R-square with different value of *d* when *D* is equal to 10 mm. R-square has a maximum value of 0.98 when *d* is equal to 11.4. The left of Figure 12a gives the temperature curves of the experiment and best matched function curve. It is worth noting that the two curves do not have good match in the platform zone of temperature curves due to the uncontrollable effect of noise in infrared thermal image. For the rest of the defect size, the determination process of corresponding *d* is the same as that of *D* = 10 mm. When the maximal degree of fitting in function curves is achieved, *d* can be confirmed for each temperature curve. Figure 13 shows that the relationship between *D*, and *d* is directly proportional, approximatively. The linear fitting formula can be written as follows:

$$D/l_{\mathrm{m}} = 0.49323 + 1.0812d/l_{\mathrm{p}} \tag{9}$$

where l_{m} is equal to 1 mm and l_{p} is one pixel. Therefore, based on the directly proportional relationship between *D* and *d*, quantitative analysis of defect size by surface temperature at the defective region is feasible.

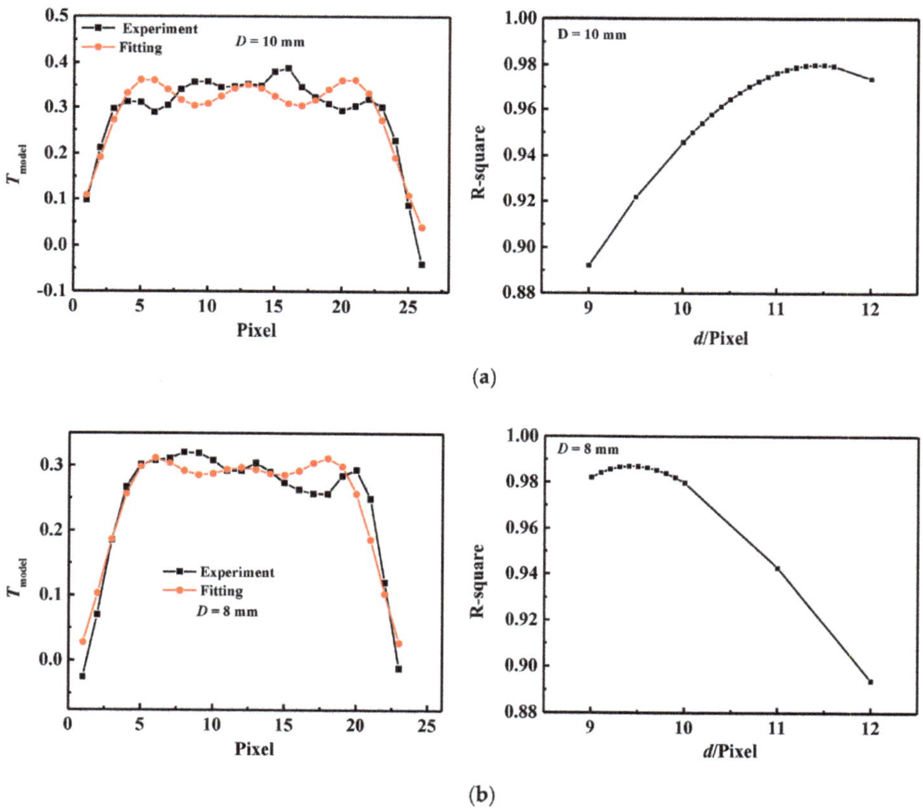

Figure 12. *Cont.*

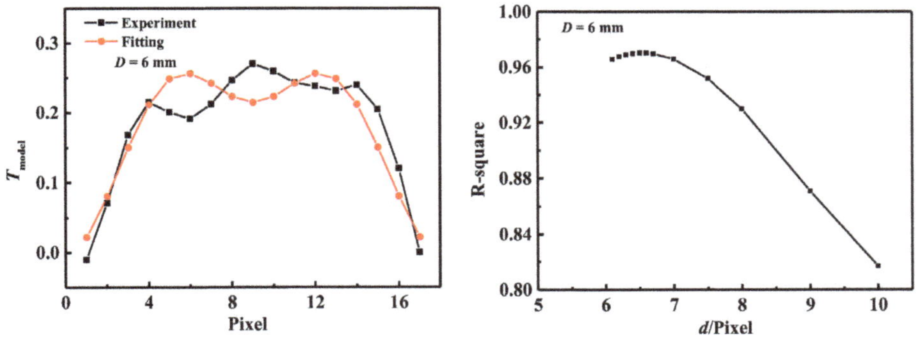

(c)

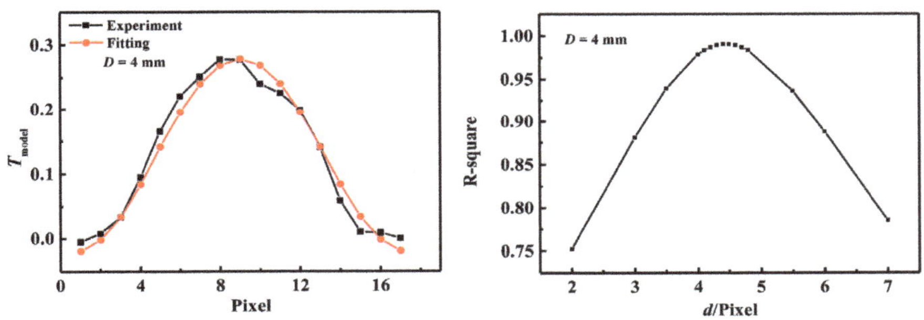

(d)

Figure 12. Temperature curve along the center of defect after background processing and fitted function curve with best degree of fitting (left). Degree of fitting of function curve vs. different value of *d* (right): (**a**) $D = 10$ mm; (**b**) $D = 8$ mm; (**c**) $D = 6$ mm and (**d**) $D = 4$ mm.

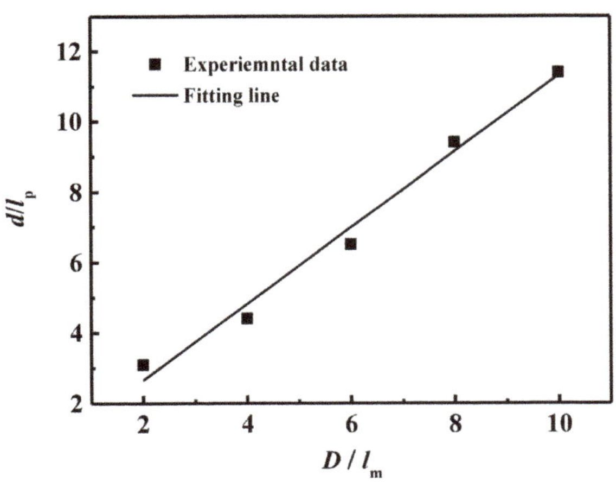

Figure 13. The defect size vs. the value of model parameter *d*.

Now, the defect size in the sample 2 can be estimated by using the aforementioned image manipulation and Equation (9). Firstly, the thermal image of defect in the sample 2 is dealt with in three processes: subtracting background image, PCA, and subtracting fitting background. Secondly, the temperature curve in the line across the center of the three flaws is extracted. The parameter d is determined by edited arithmetic fitting the Equation (8) with the extracted temperature curve. At last, the predicted defect diameter in sample 2 is obtained by Equation (9). Table 3 gives the predicted diameter and actual diameter of defect in sample 2. For the defect with diameters of 3 mm, 5 mm, and 7 mm, the predicted defect diameters are 3.1509 mm, 4.7232 mm, and 7.2205 mm. It should be noted that the predicted errors are roughly distributed in the range of 3~6%, which are not affected by defect diameter. Table 4 gives the measuring error of the interfacial defect size in TBCs predicted by the method of Guo et al. [27] is larger than 12%. The error sharply increases with the decrease of the size of defect. When the defect diameter is 3 mm, the error reaches 25%. Compared with the results obtained by Guo et al. [27], the proposed model in this work can give a more accurately evaluation of the interfacial defect size in TBCs by the collected thermal image.

Table 3. The predicted diameter and actual diameter of defect in this work.

The Actual Defect Diameter (mm)	The Predicted Defect Diameter (mm)	Error
3	3.1509	5%
5	4.7232	5.5%
7	7.2205	3.2%

Table 4. The predicted diameter and actual diameter of the prepared interfacial defect in TBCs measured by the method of Guo et al. [27].

The Actual Defect Diameter (mm)	The Predicted Defect Diameter (mm)	Error
3	2.53	25%
5	4.55	15.4%
7	6.52	12%

Ideally, the proposed method in this work can give a more accurate result for the diameter of round interfacial defects when enough training data are used to determine the parameter in Equation (9). Due to the inevitable effect of noise in infrared thermal images, this method may fail when the defect size is too small to identify. At this time, the improvement on the resolution of infrared cameras can break this limit. Furthermore, the size of interfacial defect cannot be determined directly by the proposed method when the defect shape is not regular. This is another limitation of this evaluation method. In order to solve this problem, a simple extension of the proposed method is presented. As shown in Figure 14, a defect with irregular shape is divided into different areas by five black lines. The distance between two adjacent lines is same, which is denoted by M. The points of intersection between the defect profile and these lines are represented by solid black dots. The distance between the solid dot A and B is determined by the proposed method in this work (Note that the distance between the solid dot A and B corresponds to diameter D of round defect). The same operation is used to determine the distance between the solid dots in the rest of the lines. After that, the solid dots are connected by red lines, then a red broken line that replaces the defect profile is extracted (see Figure 14). Here, we use the area surrounded by red broken line to estimate the defect area, which can be divided into four trapezoidal regions by the five black lines. The lengths of the upper side and lower side for each of trapezoidal regions are determined by the proposed method in this work, then the area surrounded by red broken line can be obtained eventually. It is worth noting that the red broken line will be closer to the defect profile when more black lines are used to split the defect area. Therefore, a more accurate assessment of the area size of defect with irregular shape can be obtained, as shown in Figure 15.

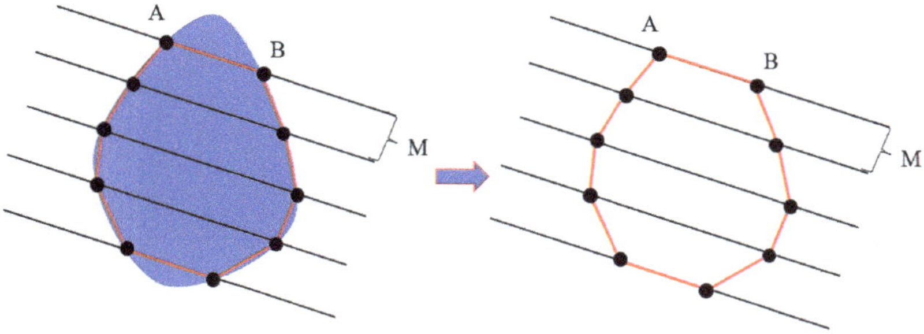

Figure 14. Schematic diagram of irregular defect. Five equidistant black lines are used to divide the area of irregular defect. A red broken line is extracted to replace the irregular defect profile.

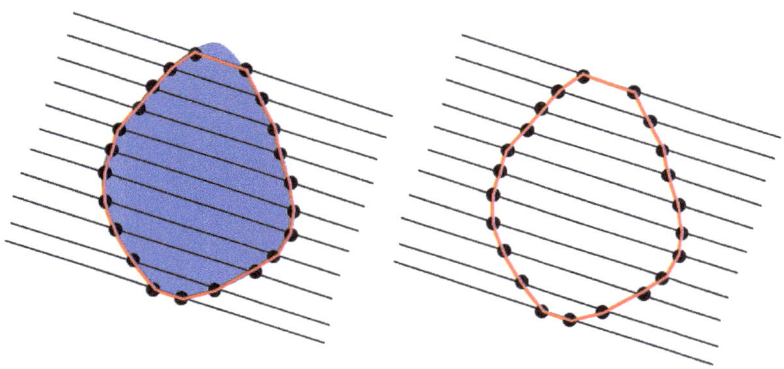

Figure 15. Schematic diagram of irregular defect. Eleven equidistant black lines are used to divide the area of irregular defect. A red broken line is extracted to replace the irregular defect profile.

4. Conclusions

In this work, the size of interface defects in TBCs is evaluated quantitatively by LPT. PCA and the method of background subtraction is proposed for improving the contrast of defects in collected infrared thermal image. The shapes of extracted temperature curves in the defective region change from platform to arch with the decrease in defect diameter. Based on the change rule of the shapes, an algorithm is edited to find the best matching result between the profile of interface defect in thermal images and the predicted shape of the model, and the interface defect size can be determined. For a defect with diameters of 3 mm, 5 mm, and 7 mm, the predicted defect diameters are 3.1509 mm, 4.7232 mm, and 7.2205 mm. The predicted errors are roughly distributed in the range of 3%~6%, which are not affected by the defect diameter. Furthermore, a simple extension of the proposed method to evaluate the size of the irregular shape defect is presented. A reconstructed broken line is presented to replace the irregular defect profile, and the area surrounded by the broken line can be estimated by the proposed method. Ideally, the broken line can be infinitely close to the defect profile when sufficient lines are used to split the defect area.

Author Contributions: Conceptualization, G.X. and W.Z.; Formal analysis, G.X. and G.L.; Funding acquisition, W.Z.; Investigation, J.W., G.X. and J.G.; Methodology, J.W. and W.Z.; Resources, J.G. and Z.M.; Software, J.W., G.L. and J.G.; Supervision, W.Z.; Validation, G.L. and Z.M.; Writing—original draft, J.W. and G.X.; Writing—review & editing, J.G. and W.Z. All authors have read and agreed to the published version of the manuscript.

Funding: This work was supported by the Science and Technology Innovation Program of Hunan Province (Grant No. 2022RC1082), the Natural Science Foundation of Hunan Province (Grant No. 2020JJ3031), the Scientific Research Foundation of Hunan Provincial Education Department (Grant

No. 21A0120) and Postgraduate Scientific Research Innovation Project of Hunan Province (Grant No. XDCX2021B145).

Institutional Review Board Statement: Not applicable.

Informed Consent Statement: Not applicable.

Data Availability Statement: Not applicable.

Conflicts of Interest: The authors declare that they have no known competing financial interests or personal relationships that could have appeared to influence the work reported in this paper.

References

1. Darolia, R. Thermal barrier coatings technology: Critical review, progress update, remaining challenges and prospects. *Int. Mater. Rev.* **2013**, *58*, 315–348. [CrossRef]
2. Vo, D.T.; Mai, T.D.; Kim, B.; Ryu, J. Numerical study on the influence of coolant temperature, pressure, and thermal barrier coating thickness on heat transfer in high-pressure blades. *Int. J. Heat Mass Transf.* **2022**, *189*, 122715. [CrossRef]
3. Pan, W.; Phillpot, S.R.; Wan, C.; Chernatynskiy, A.; Qu, Z. Low thermal conductivity oxides. *MRS Bull.* **2012**, *37*, 917–922. [CrossRef]
4. Feuerstein, A.; Knapp, J.; Taylor, T.; Ashary, A.; Bolcavage, A.; Hitchman, N. Technical and economical aspects of current thermal barrier coating systems for gas turbine engines by thermal spray and EBPVD: A review. *J. Therm. Spray Technol.* **2008**, *17*, 199–213. [CrossRef]
5. Padture, N. Advanced structural ceramics in aerospace propulsion. *Nat. Mater.* **2016**, *15*, 804. [CrossRef]
6. Benini, B. *Progress in Gas Turbine Performance*; BoD-Books on Demand; IntechOpen: London, UK, 2013.
7. Jonnalagadda, K.P.; Eriksson, R.; Li, X.-H.; Peng, R.L. Thermal barrier coatings: Life model development and validation. *Surf. Coat. Technol.* **2019**, *362*, 293–301. [CrossRef]
8. Padture, N.P.; Gell, M.; Jordan, E.H. Thermal barrier coatings for gas-turbine engine applications. *Science* **2002**, *296*, 280–284. [CrossRef]
9. Zhen, Z.; Wang, X.; Shen, Z.; Mu, R.; He, L.; Xu, Z. Thermal cycling behavior of EB-PVD rare earth oxides co-doping ZrO_2-based thermal barrier coatings. *Ceram. Int.* **2021**, *47*, 23101–23109. [CrossRef]
10. Wei, Z.-Y.; Cai, H.-N.; Zhao, S.-D. Study on spalling mechanism of APS thermal barrier coatings considering surface vertical crack evolution affected by surrounding cracks. *Ceram. Int.* **2022**, *48*, 11445–11455. [CrossRef]
11. Choi, S.R.; Hutchinson, J.W.; Evans, A.G. Delamination of multilayer thermal barrier coatings. *Mech. Mater.* **1999**, *31*, 431–447. [CrossRef]
12. Mao, W.; Dai, C.; Zhou, Y.; Liu, Q. An experimental investigation on thermomechanical buckling delamination failure characteristic of air plasma sprayed thermal barrier coating. *Surf. Coat. Technol.* **2007**, *201*, 6217–6227. [CrossRef]
13. Ma, Z.; Sun, L.; Chen, Y.; Lin, L. Ultrasonic prediction of thermal barrier coating porosity through multiscale-characteristic-based Gaussian process regression algorithm. *Appl. Acoust.* **2022**, *195*, 108831. [CrossRef]
14. Yang, X.; Wei, L.; Li, J.; Zhang, B.; Wang, S.; Guo, H. Microstructural evolution of plasma spray physical vapor deposited thermal barrier coatings at 1150 °C studied by impedance spectroscopy. *Ceram. Int.* **2018**, *44*, 10797–10805. [CrossRef]
15. Ahmadian, S.; Browning, A.; Jordan, E.H. Three-dimensional X-ray micro-computed tomography of cracks in a furnace cycled air plasma sprayed thermal barrier coating. *Scr. Mater.* **2015**, *97*, 13–16. [CrossRef]
16. Zhu, W.; Cai, X.; Yang, L.; Xia, J.; Zhou, Y.; Pi, Z. The evolution of pores in thermal barrier coatings under volcanic ash corrosion using X-ray computed tomography. *Surf. Coat. Technol.* **2019**, *357*, 372–378. [CrossRef]
17. Li, Y.; Song, Y.; Yang, Z.; Xie, X. Use of line laser scanning thermography for the defect detection and evaluation of composite material. *Sci. Eng. Compos. Mater.* **2022**, *29*, 74–83. [CrossRef]
18. Zhu, J.; Mao, Z.; Wu, D.; Zhou, J.; Jiao, D.; Shi, W.; Zhu, W.; Liu, Z. Progress and trends in non-destructive testing for thermal barrier coatings based on infrared thermography: A review. *J. Nondestruct. Eval.* **2022**, *41*, 1–26. [CrossRef]
19. Cernuschi, F.; Bison, P. Thirty Years of Thermal Barrier Coatings (TBC), Photothermal and thermographic techniques: Best practices and lessons learned. *J. Therm. Spray Technol.* **2022**, *31*, 716–744. [CrossRef]
20. Kessler, S.S.; Spearing, S.M.; Soutis, C. Damage detection in composite materials using Lamb wave methods. *Smart Mater. Struct.* **2002**, *11*, 269. [CrossRef]
21. Newaz, G.; Chen, X. Progressive damage assessment in thermal barrier coatings using thermal wave imaging technique. *Surf. Coat. Technol.* **2005**, *190*, 7–14. [CrossRef]
22. Unnikrishnakurup, S.; Dash, J.; Ray, S.; Pesala, B.; Balasubramaniam, K. Nondestructive evaluation of thermal barrier coating thickness degradation using pulsed IR thermography and THz-TDS measurements: A comparative Study. *NDT E Int.* **2020**, *116*, 102367. [CrossRef]
23. Kumar, M.M.; Pavan, A.; Markandeya, R.; Singh, K. Defect assessment in gas turbine blade coatings using non-contact thermography. *Mater. Today: Proc.* **2021**, *44*, 4414–4420.
24. Shi, L.; Long, Y.; Wang, Y.; Chen, X.; Zhao, Q. Online nondestructive evaluation of TBC crack using infrared thermography. *Meas. Sci. Technol.* **2021**, *32*, 115008. [CrossRef]

25. Zhu, W.; Li, Z.Y.; Yang, L.; Zhou, Y.C.; Wei, J.F. Real-time detection of CMAS corrosion failure in APS thermal barrier coatings under thermal shock. *Exp. Mech.* **2020**, *60*, 775–785. [CrossRef]
26. Zhuo, L.; Yang, X.; Zhu, J.; Huang, Z.; Chao, J.; Xie, W. Size determination of interior defects by reconstruction of subsurface virtual heat flux for step heating thermography. *NDT E Int.* **2022**, *133*, 102734. [CrossRef]
27. Guo, W.; Dong, L.; Wang, H.; Feng, F.; Xing, Z.; Ma, R.; Shao, H.; Gao, Z.; Wang, B.; Yang, J. Size estimation of coating disbonds using the first derivative images in pulsed thermography. *Infrared Phys. Technol.* **2020**, *104*, 103106. [CrossRef]
28. Rajic, N. Principal component thermography for flaw contrast enhancement and flaw depth characterization in composite structures. *Compos. Struct.* **2002**, *58*, 521–528. [CrossRef]
29. Balageas, D.L.; Roche, J.-M.; Leroy, F.-H.; Liu, W.-M.; Gorbach, A.M. The thermographic signal reconstruction method: A powerful tool for the enhancement of transient thermographic images. *Biocybern. Biomed. Eng.* **2015**, *35*, 1–9. [CrossRef]
30. Marinetti, S.; Grinzato, E.; Bison, P.; Bozzi, E.; Chimenti, M.; Pieri, G.; Salvetti, O. Salvetti, Statistical analysis of IR thermographic sequences by PCA. *Infrared Phys. Technol.* **2004**, *46*, 85–91. [CrossRef]
31. Ibarra-Castanedo, C.; González, D.; Klein, M.; Pilla, M.; Vallerand, S.; Maldague, X. Infrared image processing and data analysis. *Infrared Phys. Technol.* **2004**, *46*, 75–83. [CrossRef]
32. Wang, Z.; Zhu, J.; Tian, G.; Ciampa, F. Ciampa, Comparative analysis of eddy current pulsed thermography and long pulse thermography for damage detection in metals and composites. *NDT E Int.* **2019**, *107*, 102155. [CrossRef]

coatings

MDPI

Review

The Green Lubricant Coatings Deposited by Physical Vapor Deposition for Demanding Tribological Applications: A Review

Fanlin Kong [1,†], Jing Luan [1,2,3,*,†], Fuxiang Xie [4], Zhijie Zhang [1,*], Manuel Evaristo [2,3] and Albano Cavaleiro [2,3]

1 School of Materials Science and Engineering, Jiangsu University of Science and Technology, Mengxi Road 2, Zhenjiang 212003, China
2 Centre for Mechanical Engineering, Materials and Processes (CEMMPRE), Department of Mechanical Engineering, University of Coimbra, Rua Luís Reis Santos, 3030-788 Coimbra, Portugal; manuel.evaristo@dem.uc.pt (M.E.)
3 Advanced Production & Intelligent Systems Associated Laboratory (ARISE), Department of Mechanical Engineering, University of Coimbra, Rua Luís Reis Santos, 3030-788 Coimbra, Portugal
4 Laboratory for Sustainable Surface Engineering of Agricultural Machinery Systems, School of Machinery and Automation, Weifang University, Dongfeng East Street 5147, Weifang 261061, China
* Correspondence: uc2023140356@student.uc.pt (J.L.); zhangzhijie860109@163.com (Z.Z.)
† These authors contributed equally to this work.

Abstract: The emergence of nanotechnology and surface engineering techniques provides new opportunities for designing self-lubricant coatings with enhanced properties. In recent years, green coating technologies have played a vital role in environmental preservation. This article mainly reviews five typical types of self-lubricant coatings including MoN coatings, VN coatings, WN coatings and TMN (Transition Metal Nitride) soft-metal coatings, and DLC (Diamond-like Carbon) with lubricant agents deposited by PVD (Physical Vapor Deposition) for the demanding tribological applications, which is the latest research into the green lubricant coatings. Furthermore, it is of great significance for designing the green self-lubricant coatings to adapt the demanding tribological applications to meet the industrial requirements.

Keywords: green; self-lubricant; PVD; tribological applications

1. Introduction

In recent decades, there has been a growing focus on emerging green technologies that aim to utilize natural resources effectively to minimize waste and pollution. This shift towards sustainability was catalyzed by the introduction of the 2030 Agenda UN Sustainable Development Goals (SDGs), gaining significant support from the research and development (R&D) community [1,2]. According to the SDG 7: Affordable and Clean Energy, the green lubricant coatings can reduce friction and wear in machinery, leading to increased energy efficiency and lower energy consumption. Green coating technologies are particularly notable for their emphasis on environmental conservation [3]. The design and implementation of eco-friendly coatings are playing an increasingly crucial role on a global scale. Holmberg and Erdemir [4] investigated the fact that tribological contacts account for approximately 23% of total energy consumption worldwide. Moreover, roughly 20% is expended solely to overcome friction, while the remaining 3% is allocated towards the refurbishing of worn components and providing spare equipment due to wear and associated malfunctions.

Coatings refer to layers of material applied to surfaces to alter their properties or appearance [5–9]. The self-lubricant coatings are rooted in the broader context of sustainability and environmental consciousness within the field of materials science and engineering and have been developed during recent decades for demanding tribological applications, for energy saving and friction reduction. Several studies have explored the field of self-lubricant coatings. For instance, the self-lubricating MoS_2/C sputtered coating demonstrated a characteristic amorphous structure and high hardness when different carbon contents were incorporated [10]. Additionally, the investigation delved into the intricate aspects of synthesizing, understanding the microstructure, and assessing the mechanical properties of W–S–C self-lubricant thin films, and these films were meticulously deposited using magnetron sputtering, revealing insights into their behavior and performance. [11]. Pimentel et al. [12] studied the tribological properties of W–S–C–Cr self-lubricant coatings, whereas Mauafov et al. [13] and his team explored the impact of deposition conditions on the chemical composition, structure, and mechanical properties of self-lubricant W–S–N coatings. Moreover, the inclusion of amorphous SiN_x in self-lubricant Mo_2N-Ag composite films was found to enhance their mechanical and tribological properties [14]. The study meticulously examined how the introduction of novel self-lubricating TiSiVN films influenced the intricate dynamics of topographical changes, diffusion mechanisms, and oxidation phenomena occurring at the interface between the chip and the tool during the dry machining of the Ti_6Al_4V alloy. [15]. Lastly, the investigation meticulously explored the tribological properties of the $WC_{10}Co_4Cr_4CaF_2$ wear-resistant self-lubricating coating, carefully assessing its performance under diverse test temperatures to gain comprehensive insights into its behavior across a spectrum of thermal conditions [16]. Self-lubricant coatings draw inspiration from the concept of a 'chameleon'. Previous research has explored adaptive 'chameleon' nanocomposite coatings, demonstrating excellent tribological properties across different environments [17–22]. These coatings earn their 'chameleon' moniker due to a capability for self-guided, complex adaptive behaviors, which empowers them to seamlessly acclimate to a multitude of environments and temperatures, demonstrating exceptional versatility in their responses to varying conditions.

Research focusing on green coatings has been relatively limited in the past. Gautam and colleagues [23] explored the applications of green nanomaterials in coatings. Buch and Oza [24] conducted a comparative study on the machining parameters of High-Speed Steel (HSS) cutting tools, their focus centered on tools coated with multi-walled carbon nanotubes, using an eco-conscious coating technique. Through their rigorous examination, they uncovered compelling evidence: the coated Multiwall Carbon Nanotube (MWCNT) tool surpassed its uncoated counterpart, demonstrating extended lifetime, enhanced wear resistance, increased hardness, and superior surface quality. In a recent development, a new environmentally friendly film was developed by incorporating silver nanoparticles into an amorphous SiN_x matrix via magnetron sputtering. This film demonstrated long-term lubrication capabilities at temperature cycling conditions ranging from room temperature to 500 °C [25]. Jian et al. [26] introduced a pioneering method aimed at crafting durable and environmentally sustainable coatings for safeguarding magnesium alloy substrates. This approach offers a green and eco-friendly solution specifically tailored for fabricating corrosion protection coatings on AZ91D magnesium alloy. Furthermore, Fan and his team [27] developed a brand-new green coating technology aimed at achieving the controllable release of energy from nitrocellulose-based propellants. Their innovative approach provides a means to finely tune the spherical propellant energy release process with precision.

When evaluating tribological properties, we can consider factors such as the coefficient of friction (COF) [28], wear rate [29], wear track [30], and wear mechanism [31]. According to Vitu et al. [32], a tribolayer forms on the surface of the wear track. During friction, coatings composed of transition metal dichalcogenides (TMDs) reorient themselves [33]. This reorientation allows the material to slide more easily under shear forces, forming a shear-friendly layer with self-lubricating properties on the surface of the wear track. The presence of this tribolayer can significantly reduce the friction coefficient and wear rate, thereby improving the wear resistance and extending the service life of the coating. In this paper, we mainly discuss the tribological applications of coatings from COF, wear rate and wear track.

Green self-lubricant coatings represent an environmentally friendly surface treatment that offers lubrication to mitigate friction and wear between moving surfaces. These coatings present numerous advantages over conventional lubricants, such as oils and greases, including reduced environmental impact, enhanced efficiency, and increased durability. This work provides a comprehensive review of previous studies, categorized into five parts: MoN coatings, VN coatings, WN coatings, TMN soft-metal coatings, and DLC with lubricant agents. Therefore, the research on green lubricant coatings deposited by PVD for demanding tribological applications holds significant importance.

2. Physical Vapor Deposition Techniques

Multilayered films are typically fabricated through diverse methods, such as physical vapor deposition (PVD) and chemical vapor deposition (CVD). PVD, a widely utilized manufacturing method, is employed in producing thin films adapted to diverse applications, encompassing optical, mechanical, electrical, acoustic, or chemical functionalities [34]. Among PVD techniques, magnetron sputtering and arc ion plating are commonly employed methods [35]. Meanwhile, CVD is a vacuum deposition technique renowned for generating high-quality, high-performance solid materials. It finds extensive use in the semiconductor industry for thin film production [36]. Additionally, electroplating, alternatively termed electrochemical deposition, is achieved by reducing metal cations in a solution using direct electric current [37].

In this study, emphasis is placed on coatings fabricated through PVD techniques. Figure 1 gives the main types of physical vapor deposition. In this part, we will highlight two types of PVD, magnetron sputtering and arc ion plating techniques.

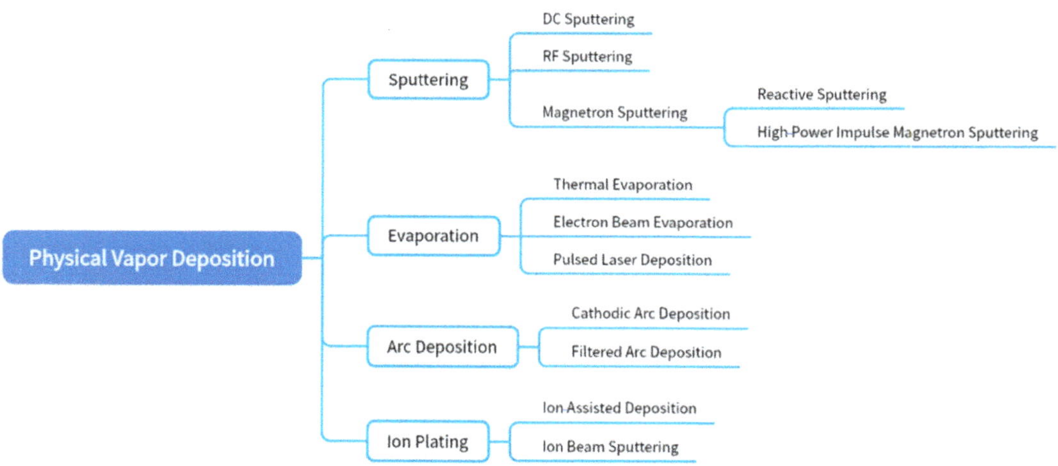

Figure 1. The main types of physical vapor deposition.

2.1. Magnetron Sputtering

Magnetron sputtering, a prominent example of PVD, offers precise control over film composition by adjusting the conditions [38,39]. Typically, sputtering equipment consists of essential components including vacuum-pumping equipment, the systems of water-cooling, sputtering-gas supply, power supply and mechanisms for positioning both substrates and targets. The target functions as the supplier of the film-forming material, while the substrate gathers the sputtered material to produce the film. During magnetron sputtering, an electric field is created through the application of voltage between the cathode (target) and the anode (either an independent electrode or the grounded wall of the vacuum chamber) [40]. Typically, the cathode target is linked to a power supply with a negative charge (Direct Current: DC or Radio Frequency: RF). The substrate is positioned on a substrate holder, which offers the flexibility to either be grounded, left floating, or biased, as needed [41]. Permanent magnets or electromagnets are strategically positioned adjacent to the target, generating a magnetic field that extends above the target surface through the alignment of magnetic field lines [42,43].

2.2. Arc Ion Plating

As an advanced surface treatment method, arc ion-plating leverages arc-discharge plasma technology to precisely deposit metal ions onto a substrate surface. This meticulously controlled process culminates in the creation of a thin film, enhancing the substrate's properties. This process initiates with high-energy arc discharge, creating plasma that evaporates the metal source. Subsequently, the plasma interacts with ions in the vapor, facilitated by ions from the reaction gas, inducing ionization and the consequent deposition onto the substrate surface [44–46]. Arc ion plating is capable of producing high-quality multilayered nanofilms.

Magnetron sputtering excels in producing coatings with excellent adhesion and smooth surfaces, which are crucial for applications like optical lenses, where even minor surface imperfections can significantly impact performance. This technology is also advantageous for creating multilayered structures, often required in advanced electronic devices. The precise control over deposition parameters allows for fine-tuning the film's properties, such as thickness, composition, and crystal structure, to meet specific application needs.

Arc ion plating is particularly beneficial for producing coatings for high-performance applications such as aerospace components, automotive parts, and medical devices. The method's ability to produce coatings with superior mechanical properties, such as enhanced hardness and resistance to oxidation and corrosion, makes it invaluable in environments demanding the utmost reliability and longevity. Additionally, the process can create aesthetically pleasing finishes, a significant advantage for decorative applications in consumer goods and architectural elements.

In the applications, including those demanding high-quality, low-temperature, and uniform coatings, magnetron sputtering is more suitable for green lubricant coatings. Conversely, in situations where high hardness and rapid deposition are essential, arc ion plating proves more advantageous. Green lubricating coatings deposited by PVD are typically designed to reduce the coefficient of friction and reduce energy losses [47,48]. Magnetron-sputtering and arc ion-plating technologies provide a high-quality, uniform and dense substrate for these coatings, helping to ensure the durability and efficiency of the lubricating coating.

3. Self-Lubricant Coatings for Tribological Applications

3.1. MoN Coatings

Molybdenum nitride (MoN) films, a representative type of transition metal nitride (TMN), can be applied for diverse applications due to their outstanding mechanical performance and the easy formation of self-lubricating phases such as Mo_4O_{11} and MoO_3. Consequently, MoN-based films have been the subject of extensive study [49–51].

The MoN multilayer, marked by a combination of crystalline and amorphous traits during deposition, displayed robust interfaces, ensuring impeccable adhesion and remarkable crack resistance [52]. Conversely, the M_{CP} film demonstrated a capability to suppress the preferred orientation through the strategic stacking of crystalline layers in alternation. Similar effects were observed in the M_{CA} film, which combined crystalline and amorphous building layers. However, a slight overgrowth of grains into adjacent layers was evident in both films. In the M_{PA} film, a prominent issue arose due to the excessive overgrowth of crystalline grains spanning multiple layers. This phenomenon led to extensive grain extrusion in the later stages of deposition, ultimately resulting in the disruption of intact layer interfaces. Molybdenum nitrides have the capability to crystallize into diverse crystal structures, such as γ-Mo_2N (cubic) and δ-MoN (hexagonal). Jauberteau and his team [53] demonstrated that the stable, non-stoichiometric γ-$Mo_2N1 \pm x$ phase, with a NaCl-B1-type cubic structure, displays favorable catalytic characteristics.

Deposited by the PVD process, molybdenum nitrides and carbides exhibit numerous distinctive physical and mechanical attributes. These include high hardness, low average friction coefficients, excellent adherence to steel substrates, and low electrical resistivity, making them highly desirable for various applications [54–57]. Abboudi et al. [58] conducted a study to analyze how variations in nitrogen content and film thickness affect the structural and tribo-mechanical properties of MoN thin films deposited using reactive sputtering. In the MoN film, they observed a significant compressive residual stress of −5.7 GPa with a thickness of 0.3 μm, which progressively decreased as the film thickness increased and the friction coefficient of MoN films, tested with Si_3N_4 balls, varied within the range of 0.55 to 0.93. The Mo_xN coating, deposited by high-power impulse magnetron sputtering with a peak current of 260 A, stands out for its notably low friction coefficient of 0.28, while the same coating applied with a peak current of 300 A showcases exceptional wear resistance, characterized by achieving the lowest wear rate, observed at 5×10^{-8} $mm^3/(Nm)$ [59].

The tribological properties of Mo_2N films are profoundly influenced by temperature variations [60]. As the temperature rises from 25 °C to 200 °C, there is a reduction in moisture absorption and the weight fraction of MoO_3. Consequently, this causes an elevation in the average friction coefficient and a reduced wear rate, with the lowest average friction coefficient being 0.29 (550 °C). Gassner and his team found that by suppressing the formation of the volatile MoO_3 phase, such as limiting the maximum testing temperatures to 500 °C, Mo–N coatings demonstrate excellent lubricious properties and this leads to significantly lower friction coefficients of 0.4 [61]. Furthermore, heat treatment has been shown to enhance the mechanical and tribological performance of MoN-based films and the COF of the MoN film stabilized at approximately 0.34 before annealing; it then experienced a slight decrease and stabilized at about 0.30, after annealing [62]. As for the wear track, MoN film has large scratches, while the wear scar morphology shows itself to be smoother than before, after annealing. MoN/VN multilayer films have demonstrated improved tribological properties, particularly at room temperature (25 °C) and 700 °C, due to the occurrence of sliding wear and the formation of self-lubricating oxide films [63]. The MoN/VN multilayer films with Λ = 22.66 nm displayed exceptional performance, with the lowest recorded average friction coefficient of 0.29 and a minimal wear rate of 1.37×10^{-6} mm^3/Nm at 700 °C. MoN coatings deposited by ion beam-assisted deposition achieve the lowest COF of 0.27 at a deposition temperature of 600 °C [64].

Apart from magnetron sputtering, MoN coatings are also studied by other methods. MoN coatings via CVD (chemical vapor deposition) exhibited exceptional properties [65]. MoN coatings were prepared by the cathodic arc evaporation method and the specific wear rates varied among the coatings and ranged between (1×10^{-7} and 2.1×10^{-6}) mm^3 N^{-1} m^{-1} [66]. MoN coatings for pistons in a diesel engine were analyzed by Hazar [67], who found that the MoN-coated piston exhibited reduced deformation and fewer wear-induced scratches in comparison to the uncoated piston; the elevated surface hardness of the MoN-coated surface effectively minimizes wear traces.

Later, the researchers endeavored to design the multilayer coatings to enhance their mechanical or tribological properties. According to Guojun Zhang [68], the MoN_x/SiN_x multilayer coatings show a friction coefficient of around 0.53. For MoN/MoSiN coatings, it was found that the substantial volumetric ratio of c-MoN to a-MoSiN, along with the discontinuities within the crystalline phases, were fundamental factors contributing to their advantageous mechanical properties [69]. Moreover, self-lubricant nano-multilayer coatings such as TiN/MoN and TiAlN/MoN, produced using a closed-field unbalanced magnetron sputtering system on high-speed steel drills, have been shown to enhance tool durability and surface quality during drilling operations [70]. Table 1 shows part of the mechanical/tribological properties of MoN coatings.

Table 1. Mechanical/Tribological properties of MoN coatings.

System	Techniques	Mechanical/Tribological Properties	References
MoN	Reactive magnetron sputtering	Hardness: ~9.5 GPa–~35 GPa; coefficient of friction: 0.55	[58]
Mo_xN	High-power impulse magnetron sputtering	Coefficient of friction: 0.28; wear rate: 5×10^{-8} $mm^3/(Nm)$	[59]
Mo_2N	Reactive magnetron sputtering	Coefficient of friction: 0.29~0.53; wear rate: 2.1×10^{-6} mm^3/Nmm~ 5.3×10^{-7} mm^3/Nmm	[60]
MoN_x	Reactive magnetron sputtering	Hardness: ~10.6 GPa–~31.4 GPa; coefficient of friction: 0.5~0.93	[71]
MoN	Vacuum arc plasma-assisted	Hardness: ~36 GPa; wear rate: 2.08×10^{-7} mm^3N^{-1} m^{-1}	[72]
MoN_x	DC magnetron sputtering	Hardness: ~27 GPa	[73]
MoN_x	DC magnetron sputtering	Hardness: ~23 Gpa	[74]
MoN	High-power impulse magnetron sputtering	Hardness: >~22 GPa; elastic modulus: >300 GPa; coefficient of friction: 0.7	[75]
MoN_x	DC magnetron sputtering	Hardness: ~18 GPa–~24 GPa; coefficient of friction: 0.04~0.08	[76]
MoN_x	DC pulsed magnetron sputtering	Hardness: ~16 GPa; elastic modulus: ~180 GPa	[77]
MoN	Vacuum arc plasma-assisted	Coefficient of friction: 0.3~0.7	[78]

The previous studies discovered that MoN coatings exhibit significant compressive stress and that their friction coefficients vary, based on nitrogen content and thickness. Various deposition methods, such as high-power pulsed magnetron sputtering, ion beam-assisted deposition, and vacuum arc deposition, influence the tribological and mechanical properties of MoN coatings. MoN-based multilayer coatings demonstrate excellent self-lubricating properties at high temperatures, with their performance further enhanced by heat treatment. Overall, MoN coatings show great potential in enhancing the wear resistance of mechanical equipment and reducing friction losses.

In the future, we will further investigate the self-lubricating properties of MoN-based multilayer coatings under high-temperature conditions. We will also explore methods to enhance their high-temperature performance by incorporating other lubricating materials or chemical elements. Additionally, we plan to develop composite coating systems that include MoN, such as those combining MoN with other transition metal nitrides or carbides. This approach aims to further improve the overall performance of the coatings and enable them to meet more diversified application requirements. The tribological properties of MoN coatings are of great significance in engineering and applications. MoN coatings usually have high hardness and excellent wear resistance, which can effectively reduce friction and surface wear. This is crucial to improving the working efficiency and extending the service life of mechanical parts, especially in equipment with a high-load, high-speed or harsh environment.

3.2. VN Coatings

As one of the most promising options for tribological hard coatings, vanadium nitride (VN) coating has attracted considerable interest; it is susceptible to oxidation at elevated temperatures and can melt, leading to the creation of a lubricant layer [79,80].

The integration of transition metal vanadium (V) into nitride coatings catalyzes the emergence of Magnéli oxide lubricant phases, characterized by their inherently sheared planes, in the context of machining or sliding applications. These phases serve as efficacious agents in reducing both the coefficient of friction and the wear rate of the coatings, thereby bolstering their durability and operational efficiency [81,82]. Optimizing the nitrogen flow rate to 2.5 sccm yields a VN film characterized by exceptional properties: elevated hardness, diminished electrical resistivity, and minimal residual stress, and this fine-tuned parameter underscores the film's superior performance and stability [83]. The fracture toughness of VN coatings is dependent on texture, and specimens displaying random texture and (200) texture manifest Gc values of 11.2 and 19.2 J/m^2, respectively [84]. However, Ge and his team [85] enhanced the orientation of constituent columns and discovered that suppressing fracture-dominated wear mechanisms in dense VN coatings can significantly reduce wear rates. Moreover, denser VN coatings were obtained with higher hardness (22.9 GPa) at 0.29 Pa, through the influence of nitrogen partial pressure and substrate bias on the mechanical properties of VN coatings [86].

There is some research focused on tribological properties and a wide range of temperatures. The tribological properties of VN films are greatly improved after thermal treatment, with the VN film featuring a VO_2 thin pre-oxide layer exhibiting a significantly lower coefficient of friction of approximately 0.46 and a wear rate of $(2.27 \pm 0.14) \times 10^{-15}$ m^3 N^{-1} m^{-1}, compared to the VN film without the pre-oxide layer, highlighting the critical role of the pre-oxide layer in enhancing the film's properties [87]. VN coatings were deposited by reactive unbalanced magnetron sputtering, and exhibited self-lubricating properties due to the liquid/solid lubrication, particularly effective at high temperatures, resulting in low friction coefficients during dry-sliding against various counterpart materials [88]. There are studies which have also focused on the performance of VN coatings across a wide temperature range. The width of the wear track for the VN coating initially increases up to 300 °C, after which it decreases until reaching 500 °C [89]. However, at 700 °C, significant delamination and the presence of abundant wear debris can be observed on the wear tracks of the VN coatings. Regarding the COF, it shows a tendency to decline for the VN coatings until 500 °C, reaching its lowest value at (0.36). This decrease can be attributed to the formation of a lubricious phase, likely V_2O_5. However, the COF remains relatively constant at higher-friction temperatures (700 °C). Zhou et al. [90] pointed out that the melting point of the V_2O_5 phase is approximately 690 °C, which is significantly lower than the friction temperature of 700 °C, resulting in a notable decline in the concentration of the V_2O_5 phase as the temperature reaches 700 °C. As the friction temperature rises, the wear rates of VN coatings initially increase, reaching peak values around 8.732×10^{-6} mm^3/Nm, at approximately 500 °C. This initial rise in wear rate is caused by the evaporation or removal of a condensed water-vapor film present on the sliding surface, which provides a lubricating effect [91]. As temperatures continue to increase beyond 500 °C, extensive surface oxidation occurs due to prolonged exposure to high temperatures and sliding wear [92]. In Fateh's study [93], the friction coefficient of the VN coating continues to decrease above 400 °C and reaches 0.25 at 700 °C. This decrease is attributed to the formation and subsequent melting of V_2O_5, which facilitates liquid lubrication. Moreover, the VN coating exhibits self-lubricating properties at elevated temperatures.

As for the VN-based coatings, the cracking of V_xN coatings induced by scratching is primarily confined to larger droplets exhibiting weak interfacial bonding with the surrounding coating matrix [94]. When comparing VN/VCN coatings, researchers noted that the multilayer VN/C coating displayed markedly superior properties in both wear resistance and corrosion resistance compared to the monolayer VN coatings [95]. Furthermore, the VCN coating exhibited consistently lower friction coefficients and reduced wear rates across a spectrum of temperatures when compared directly with the VN coating [96,97]. These findings underscore the advantages of multilayer structures and the beneficial effects of carbon incorporation in enhancing the performance of these coatings in demanding tribological applications. Cai et al. [98] made a study of tribological properties of VN-based coatings and obtained friction coefficients for the VCN, VAlN, VCN and VAlCN/Ag coatings which decreased significantly to 0.3, 0.37/0.42, and 0.25, respectively, at 550 °C. This notable reduction can be attributed to the formation of the V_2O_5 Magnéli phase, known for its low shear strength, which initiates above 500 °C [99,100]. Moreover, at elevated temperatures, silver atoms have the capacity to migrate towards the surface of the coating through grain boundaries. Upon reaching the surface, these atoms undergo reactions to form a lubricious layer of reaction products with low shear strength. This layer, deposited onto the worn surface, plays a significant role in further decreasing the friction coefficient of the VAlCN/Ag coating [101,102]. The following Table 2 shows the mechanical/tribological properties of VN coatings.

Table 2. Mechanical/Tribological properties of VN coatings.

System	Techniques	Mechanical/Tribological Properties	References
VN	Reactive magnetron sputtering	Hardness: ~25–~30 GPa; wear rate: <5 × 10^{-17} m^3/N m	[85]
VN	Reactive magnetron sputtering	Hardness: ~22.9 GPa;	[86]
VN	Arc ion plating method	Coefficient of friction: 0.23/0.46; wear rate: (11.41 ± 0.22) × 10^{-15} m^3 N^{-1} m^{-1} /(2.27 ± 0.14) × 10^{-15} m^3 N^{-1} m^{-1}	[87]
VN	Reactive magnetron sputtering	Coefficient of friction: 0.5	[88]
V_xN	Arc evaporated	Coefficient of friction: 0.15–0.20;	[94]
VN	Reactive magnetron sputtering	Hardness: ~11 GPa–~20 GPa	[103]
VN	Inductively coupled plasma (ICP)-assisted sputtering	Hardness: ~28.2 GPa	[104]
VN	Arc evaporation method	Hardness: ~19.00 ± 1.26 GPa; coefficient of friction: 0.30; wear rate: 1.55 × 10^{-6} mm^3/Nm	[105]
VN	Arc evaporation method	Hardness: ~37 GPa; wear rate: 1.7 × 10^{-6} mm^3/Nm	[106]

Based on the research findings, the tribological properties of VN coatings vary significantly with temperature changes, demonstrating notably lower friction coefficients and wear rates after heat treatment, particularly due to the beneficial impact of the pre-oxidation layer. At high temperatures, VN coatings exhibit self-lubricating properties, primarily resulting from the liquid/solid lubrication effect. The friction coefficient decreases steadily up to 500 °C, achieving a minimum value of 0.36, mainly attributable to the formation of the V_2O_5 lubricating phase. However, at 700 °C, the friction coefficient stabilizes while the wear rate peaks at 500 °C, with surface oxidation intensifying as the temperature increases further. Additionally, the previous studies reveal that VN/VCN multilayer coatings offer superior wear and corrosion resistance compared to single-layer VN coatings. VN-based coatings demonstrate significantly reduced friction coefficients and wear rates at high temperatures, with VAlCN/Ag coatings, in particular, exhibiting excellent self-lubricating properties. The lubricating layer formed by the migration of silver atoms further contributes to reducing the friction coefficient. However, the urgent technological demand lies in the development of vanadium nitride coatings that are hard and resilient yet possess self-lubricating characteristics. This advancement is crucial for mitigating mechanical friction losses and curbing energy consumption. In addition, the green multilayer self-lubricant coatings design has great prospects for the future.

3.3. *WN Coatings*

Another notable class of materials for tribological applications is tungsten nitride (WN)-based coatings, primarily utilized in machining tools [107]. Lately, a series of tungsten nitride coatings with nitrogen content ranging from 0 to 55 atomic percent were fabricated and subsequently analyzed [108]. Generally, the trend of increasing hardness was observed with higher nitrogen content, particularly evident in the α-W and β-W structures associated with lower nitrogen levels. Conversely, for the β-W_2N phase, the opposite trend was noted. Boukhris et al. [109] found that high-quality W_2N films can be achieved with a dense structure and a resistivity of approximately 200 μΩ cm.

In terms of mechanical properties, the suitability of tungsten nitride coatings, produced through High Power Impulse Magnetron Sputtering (HiPIMS), was examined for their potential application as plasma-facing materials in fusion environments [110]. Furthermore, a first-principle study was conducted to assess how ultrathin WN films behave regarding surface stability under exposure to oxygen (O_2) and water vapor (H_2O). Patra et al. [111] found that optimal conditions are required to achieve defect-free surfaces in energetically stable ultrathin films terminated with tungsten (W), thereby ensuring surfaces with remarkable mechanical strength.

As a protective layer, WN coatings, such as TiN HEC (electrode contact), safeguard PCM (phase-change memory) electrodes from oxidation during manufacturing and fatigue processes [112]. Among these coatings, the most resilient W/W–N coating, featuring 5-nm layers, demonstrates superior protective properties [113]. Furthermore, W and WN_x films were examined at temperatures reaching 500 °C, and films with $x \leq 0.20$ revealed an α-W structure and columnar microstructure, while those with $x \geq 0.27$ exhibited a β-W_2N or δ-WN structure and fine-grained microstructure [114]. In addition, the incorporation of carbon into columnar WN coatings has been shown to enhance oxidation resistance [115].

In 2007, Polar et al. investigated the tribological properties of WN coatings and found that the coefficient of friction was 0.61 (12 at.%N) and the wear rate was 0.09×10^{-6} mm^3/Nm (55 at.% N); with the increase in N content, the key factor influencing the change in the tribological behavior of W–N coatings appears to be the presence of the NaCl-type W_2N phase [116]. WN_x coatings were applied to silicon (Si) and AISI 304 stainless steel (SS) substrates using a combined HiPIMS technology and MF (mid-frequency) impulse. The $WN_{0.12}$ coating showed a low friction coefficient of 0.33, a high hardness of 31.5 GPa, and high H/E and H^3/E^2 ratios of 0.104 and 0.342 GPa [117]. These excellent properties are attributed to solution hardening, the influence of substrate bias, and the higher peak power densities achieved during the deposition process. According to Dinesh Kumara [118], the average friction coefficient for the WN coatings deposited on silicon substrates was found to be 0.3, whereas for those deposited on steel substrates, the average value was 0.21. Based on the previous studies, the lowest COF we can achieve is 0.21.

As for the multilayer WN coatings, the relation between friction coefficient and wear rate of the WN coatings under different testing temperatures was revealed, and it was found that an increase in the nitrogen content within the coatings directly corresponds to improved wear resistance at elevated temperatures [119]. It is reported that the friction coefficient of the CrN monolayer coating is 0.63, whereas the CrN/WN multilayer film with a bilayer period of 30 nm exhibits a significantly lower friction coefficient of 0.31; the enhancement in tribo-mechanical properties of CrN/WN multilayer coatings can be attributed to the presence of the superlattice structure [120]. Moreover, the combination of ZrN/WN multilayers has demonstrated high hardness, low residual stress, high elastic modulus, and good fracture resistance [121]. The addition of tungsten (W) to TiAlCrN/WN nano-multilayered coatings has been shown to improve tool life [122]. (TiZrNbHfTa)N/WN multicomponent coatings were developed and the results revealed that the (TiZrNbHfTa)N layer exhibits the formation of a simple disordered solid solution, while the WN layer demonstrates a β-W_2N phase with an fcc crystal structure. In addition, the high-entropy alloy phases display highly disordered bcc orientations along (110) and (220) planes [123]. Table 3 shows the mechanical/tribological properties of WN coatings.

Table 3. Mechanical/Tribological properties of WN coatings.

System	Techniques	Mechanical/Tribological Properties	References
WN_x	High-power impulse magnetron sputtering	Coefficient of friction: 0.33; hardness: ~31.5 GPa	[117]
WN	DC magnetron sputtering	Coefficient of friction: 0.21–0.3	[118]
WN	Reactive magnetron sputtering	Coefficient of friction: 0.61 (12 at.%N); wear rate: 0.09×10^{-6} mm^3/Nm (55 at.%N); hardness: ~40 GPa	[119]
WN_x	Reactive magnetron sputtering	Hardness: ~28 GPa	[124]
WN_x	Reactive magnetron sputtering	Hardness: ~27.7/38.7 GPa	[125]
WN	Reactive magnetron sputtering	Hardness: ~23 GPa	[126]

Previous studies have indicated a lack of research mainly focused on the tribological properties of WN coatings, with most of the research emphasizing mechanical properties or oxidation properties. Polar et al. studied the tribological properties of WN coatings and found that with the increase in nitrogen content, the friction coefficient was 0.61 (12 at.%N) and the wear rate was 0.09×10^{-6} mm^3/Nm (55 at.%N), mainly due to the presence of the NaCl-type W_2N phase. WN coatings were applied on silicon and AISI 304 stainless steel substrates using HiPIMS and MF systems. The $WN_{0.12}$ coating exhibited a low friction coefficient of 0.33, a high hardness of 31.5 GPa, and a high H/E and H^3/E^2 ratio, with these excellent properties being ascribed to solid solution strengthening, substrate bias, and the high peak-power density during deposition. Additionally, Dinesh Kumar noted that the friction coefficient of WN coatings was 0.3 on silicon substrates and 0.21 on steel substrates. Furthermore, multilayer WN coatings demonstrated that increasing nitrogen content at various temperatures can enhance high-temperature wear resistance. Hence, it is of profound significance to investigate environmentally friendly WN-based multilayer coatings deposited by PVD for demanding tribological applications. Future research on WN coatings should include that can still maintain excellent performance under high-temperature conditions, especially coating materials that can maintain low-friction and high-wear resistance at high temperatures; it should develop multilayer-structure and composite-material coatings, alternately superimpose WN with other high-performance materials (such as TiN, CrN, etc.), combine the advantages of different materials, and significantly improve the comprehensive performance of the coating, including wear resistance, corrosion resistance and self-lubricating properties. WN coatings in high-temperature applications include cutting tools for machining hardened steels, turbine components in aerospace engines, and dies and molds in metal-forming processes. Ongoing research aims to optimize WN coatings further for specific high-temperature applications, enhancing their performance and expanding their use in various industries.

3.4. TMN-SMe Coatings

Transition metal nitrides (TMNs) encompass various types, including VN, TiN, and chromium nitride (CrN), where nitrogen atoms occupy interstitial positions within the lattice metal crystal network [127]. TMN exhibits three primary structural forms: face-centered cubic (fcc), hexagonal close-packed (hcp), and simple hexagonal (hex) [128]. In particular, transition metal nitrides generally exhibit more covalent and metallic properties compared to corresponding oxides and sulfides, endowing them with superior characteristics across multiple domains [129]. These include remarkable hardness, high melting points, high Young's modulus, excellent thermal and chemical stabilities, high electrical conductivity, and wide band gaps.

There are many studies on TMN adding Ag content. The incorporation of Ag into TiN films has been identified as a substantial enhancer of their fracture toughness and critical load, markedly improving their mechanical characteristics [130]. With the introduction of

Ag (0.20 at 41.1 at.% Ag) into TiN film, the mean friction coefficient was reduced from 0.78 to 0. Nevertheless, the wear rate of TiN/Ag films showed an initial decrease, reaching its nadir at approximately 0.8 at.% Ag, with a value of about 1.3×10^{-7} $mm^3/(mm\ N)$, before subsequently increasing. The tribological characteristics of the films were notably affected by both the presence of the lubricant Ag phase and the ratios of H/E and H^3/E^2. The introduction of Ag into the ZrN matrix resulted in a different change in the average friction coefficient of the films in a wide range of temperatures. At an Ag content of 26.6 atomic percent (at.%) at room temperature, the film achieves its lowest average friction coefficient of 0.62, yet this condition simultaneously results in the highest wear rate, peaking at 1.3×10^{-7} $mm^3/N{\cdot}mm$, whereas at an elevated temperature of 600 °C, the average friction coefficient further decreases to 0.29, although the wear rate markedly increases to 2.1×10^{-6} $mm^3/N{\cdot}mm$ [131]. Significant friction reductions for CrN/Ag coating are achieved, with the friction coefficient reaching its lowest point of 0.05 during testing at a temperature of 500 °C [132]. The TaN films doped with a small amount of solute Ag atoms (1.2 at.%) demonstrate an impressive combination of properties, including higher hardness (36.1 GPa), enhanced toughness (K_IC = 1.48 $MPa{\cdot}m^{1/2}$), and superior wear resistance, as evidenced by a minimum wear rate of approximately 1.9×10^{-6} $mm^3/N{\cdot}m$ and a significantly reduced coefficient of friction of about 0.21, making them exceptionally well-suited for applications that demand both durability and low friction [133]. A VN/Ag solid-solution coating, with up to about 5 atomic percent of Ag, was created via magnetron sputtering to enhance its self-lubricating behavior in both dry-friction and oil-lubrication conditions, showing higher hardness (~24.5 GPa) and a lower coefficient of friction (~0.04) [134]. Based on Wu's [135] study, MoN/Ag coatings containing 2.2% Ag, prepared using a hybrid PVD technology involving DC and RF magnetron sputtering, exhibited outstanding tribological properties, characterized by an exceptionally low friction coefficient of 0.27 even at a high temperature of 700 °C. The addition of Ag into NbN was studied by many researchers. Hongbo Ju and Junhua Xu [136] found that the average friction coefficient and wear rate of NbN-Ag films is influenced by the Ag content; the optimal Ag content, ranging from 9.2 to 13.5 atomic percent (at.%), was determined to exhibit low average friction coefficients ranging from 0.46 to 0.40, along with minimal wear rates ranging from 1.1×10^{-8} to 1.7×10^{-8} $mm^3/(mm{\cdot}N)$ [137]. Ping Ren et al. [138] put forward a novel method to integrate a small number of Ag atoms (~1.5 at.%) into NbN film, creating a NbN/Ag solid-solution structure, resulting in increased hardness and toughness, as well as enhanced wear-resistance ability and a significant reduction in coefficient of friction. NbN films ($\sim$1.5 at.% solute Ag) exhibited a highest hardness of 28 GPa, a minimal wear rate of 4.1×10^{-9} $mm^3/(Nm)$, and a low coefficient of friction of 0.22.

Another commonly used addition is the Cu element, in TMN. The addition of Cu to TiN films has been shown to significantly enhance wear resistance, with TiN/Cu films containing 1.6 atomic percent Cu achieving a maximum hardness value of 36 GPa under a pulse bias voltage of −200 V [139]. The addition of Cu to the MoN coating significantly reduces its wear rate under both dry and oil-lubricated conditions. Specifically, the wear rates for the MoN/Cu coating are 3.2×10^{-7} $mm^3/(Nm)$ and 2.1×10^{-7} $mm^3/(Nm)$ under dry and oil-lubricated conditions, respectively. In dry conditions, the friction coefficient of the MoN coating exhibits a brief running-in period before stabilizing at approximately 0.58. Conversely, in oil-lubricated conditions, the friction coefficient of the MoN/Cu coating initially increases gradually and eventually stabilizes around 0.053 [140]. As for the NbN/Cu coatings, the COF was observed to rise with the incorporation of copper into the NbN coatings, and NbN/Cu coatings with 24.23 at.% Cu showed a high wear rate of 0.93 [141].

There are other soft metals to apply in TMN coatings. The TiN/Au coatings initially demonstrate a notably low friction coefficient ($\mu = 0.07$), which undergoes a gradual increase over the course of operation until it stabilizes at a slightly higher value of $\mu = 0.12$; this frictional behavior is attributed to intra-film sliding serving as the predominant lubrication mechanism within the TiN/Au coating [142]. Compared with the pure TiN coatings, TiN/Pt multilayer toughness was higher [143]. The following Table 4 shows the mechanical/tribological properties of TMN soft-metal coatings.

Table 4. Mechanical/Tribological properties of TMN soft-metal coatings.

Systems	Techniques	Mechanical/Tribological Properties	References
TiN/Ag	Reactive magnetron sputtering	Hardness: ~29 GPa (Ag 0.8 at.%); wear rate: 1.3×10^{-7} $mm^3/(mm.N)$(Ag 0.8 at.%)	[130]
ZrN/Ag	Reactive magnetron sputtering	Hardness: ~29 GPa; coefficient of friction: 0.62 (Ag 26.6 at.%); wear rate: 1.3×10^{-7} mm^3 $N^{-1}mm^{-1}$ (Ag 26.6 at.%)	[131]
TaN/Ag	Reactive magnetron sputtering	Hardness: ~36.1 GPa (Ag1.2 at. %); wear rate: 1.9×10^{-6} mm^3/Nm	[133]
VN/Ag	Magnetron sputtering	Hardness: ~24.5 GPa ($\leq$~5 at.% Ag); Coefficient of friction: 0.04	[134]
NbN/Ag	Reactive magnetron sputtering	Coefficient of friction: 0.4–046 (Ag 9.2–13.5 at.%); wear rate: 1.1×10^{-8}~1.7×10^{-8} $mm^3/(mm\ N)$	[136]
TiN/Cu	Reactive magnetron sputtering	Hardness: 40.4 GPa	[139]
MoN/Cu	DC magnetron sputtering	Coefficient of friction: 0.58; wear rate: 2.1×10^{-7} $mm^3(Nm)^{-1}/3.2 \times 10^{-7}$ $mm^3(Nm)^{-1}$	[140]
NbN/Cu	Reactive magnetron sputtering	Hardness: 40 GPa (Cu 1 at.%); coefficient of friction: 0.93	[141]
TiN/Au	Magnetron sputtering	Coefficient of friction: 0.12	[142]

Based on the above research, TMN presents a set of notable characteristics, including high hardness, a high melting point, a robust elastic modulus, outstanding thermochemical stability, high conductivity, and a wide bandgap. Moreover, investigations have explored the integration of elements into TMN to improve its tribological performance. This strategic introduction of diverse elements, such as Ag, Cu, and Au, among others, serves to enhance various attributes of the TMN coating. When Ag is added to TMN coatings, the friction coefficient will be different when a different content of the Ag element is added and a different temperature is used. For instance, for TiN/Ag coatings with the Ag content ranging between 0.20 and 41.1, the friction coefficient decreases significantly from 0.78 to 0.29, while the friction coefficient of CrN/Ag coatings reaches a remarkably low minimum of 0.05 at 500 °C.

When Cu is added to TMN coatings, it generally enhances the hardness and wear resistance of the TMN coating, but it may also lead to an increase in the friction coefficient and wear rate. When other elements are added to TMN coatings, such as in the case of TiN/Au films, the initial friction coefficient starts at 0.07 and gradually increases to 0.12.

Consequently, the future development we should consider is that the alternating superposition of TMN layers and lubricating layers containing soft metals can provide better friction and wear performance under high-temperature and high-load conditions. Additionally, research is being conducted to maintain the stability and lubrication effect of TMN soft-metal coatings under extreme temperature conditions. With increasingly stringent environmental regulations, the development of non-toxic and environmentally friendly TMN soft-metal coatings will become an important trend.

3.5. Lubricant Agent-Doped DLC Coatings

Diamond-like carbon (DLC) coatings were comprehensively studied for their performance in gear bending fatigue, contact fatigue, and resistance to scuffing. Classified as a metastable form of amorphous carbon, these coatings exhibit properties that bear resemblance to both diamond and graphite. These include remarkable wear resistance and a low coefficient of friction. It was found that the combination of shot peening and DLC coating remains effective in enhancing the bending fatigue performance of the composite material, by Wu and his team [144]. Marian et al. [145] demonstrated that the combination of 3D (three-dimensional) printing's high flexibility and cost efficiency with DLC's exceptional mechanical and tribological properties creates synergies that culminate in outstanding performance under dry sliding conditions. DLC films, characterized as a metastable form of amorphous carbon, sharing traits of both diamond and graphite, have attracted considerable attention due to their outstanding chemical and mechanical properties and they offer high wear resistance alongside a low friction coefficient, making them highly desirable for various applications [146–156]. The remarkable stability of DLC films in both acidic and alkaline solutions make them an excellent choice for effectively inhibiting corrosion in metallic products [157,158].

The graphene/DLC coating, applied with three layers of spraying, exhibits outstanding tribological properties under vacuum conditions, featuring an average coefficient of friction of approximately 0.08 and a low wear rate of 4.59×10^{-7} $mm^3/N{\cdot}m$ [159]. In comparison with mineral-based lubricants, synthetic saturated base esters effectively reduce friction under extreme boundary conditions [160]. Tasdemir et al. [161] discussed the fact that the organic friction modifier glycerol-mono-oleate demonstrated the ability to maintain low friction properties of the coating even at elevated temperatures, with enhanced durability. Conversely, the anti-wear additive Zinc Dialkyldithiophosphate (ZnDTP) exhibited exceptional wear protection for the ta-C coating by generating a thick pad-like tribofilm on the steel counterpart across a wide range of temperatures; however, the friction coefficient increased when the coated surfaces were rubbed in ZnDTP-additivated oil. In addition, Kuznetsova et al. [162] utilized silicon addition to fine-tune the properties of tribofilms and enhance the tribological properties of DLC coatings and found that the DLC/Si(0.8%) formulation stands out, with the highest observed coefficient of friction (COF) at 0.135, whereas the DLC/Si(10%) variant displays the lowest COF along with a specific volumetric wear rate of 0.6×10^{-3} $mm^3/N{\cdot}m$, indicating its superior performance in terms of frictional characteristics and wear resistance.

When adding Perfluoropolyether (PFPE) lubricants, in simulations involving HDI (head–disk interface) contacts, it was observed that PFPE lubricants with greater quantities and stronger functional end groups tended to generate higher friction forces and exhibit reduced head contamination due to lubricant transfer. Conversely, as temperatures increased, head contamination increased while friction forces decreased [163]. DLC/biodegradable lubricants were studied, and it was found that the properties of the coating itself were found to have a more pronounced impact on wear compared with the influence of additives in the lubricant, with the wear coefficient generally remaining around the order of 0.1×10^{-18} $m^2 N^{-1}$ across all observed cases, except in instances where coating spallation occurred [164]. As for DLC/Ti-synthetic bio-lubricant coating, it is noteworthy that the friction of DLC/Ti exceeds that of pure DLC, and there is a significant reduction in wear observed when using pure DLC compared to DLC/Ti [165]. A tungsten carbide-doped DLC coating (WC/C) deposited by PVD on bearing steel was studied, and it was found that the tribological properties using water-based lubricants can be significantly enhanced with the application of this type of DLC coating [166]. When it comes to the tribological properties of DLC/MoDTC (Molybdenum Dithiocarbamate)-containing oils, Ueda et al. [167] revealed that the presence of other surface-active additives can decrease DLC wear in PAO + Mo (Mo in Polyalphaolefin base oil) through three mechanisms: the formation of thick anti-wear tribofilms, an increase in the ratio of MoS_2:MoO_3, and competitive adsorption with other surface-active additives. According to Yoshida et al. [168], the density of the DLC film decreased as hydrogen concentration increased, which in turn increased the wear coefficient of the sample, regardless of whether MoDTC was present in the oil, and an increase in the sp^2/sp^3 ratio of the DLC also increased the wear coefficient when immersed in MoDTC-containing oil. The following Table 5 illustrates the tribological properties of lubricant agent-doped DLC coatings.

Table 5. Tribological properties of lubricant agent-doped DLC coatings.

Systems	Technique	Mechanical/Tribological Properties	Reference
W/DLC	Magnetron sputtering	Coefficient of friction: 0.2	[169]
Ni/DLC	Magnetron Sputtering	Coefficient of friction: 0.17; wear rate: 1.6×10^{-17} mm^3/Nm	[170]
Cr-DLC	Magnetron Sputtering	Coefficient of friction: 0.17; wear rate: 1.6×10^{-17} mm^3/Nm	[171]
Ti/DLC	PECVD(Plasma-Enhanced Chemical Deposition)	Coefficient of friction: 0.15 (19 at.% of Ti)	[172]
W-DLC/Ti-DLC	PECVD	Coefficient of friction: 0.21 for Ti-DLC coefficient of friction: 0.22 for W-DLC	[173]
Ag/DLC	Magnetron sputtering	Coefficient of friction: 0.16; wear rate: 14×10^{-17} mm^3/Nm (2.99 at.% of Ag)	[174]

On the basis of the previous studies, DLC coating has undergone extensive examination for its efficacy in enhancing gear bending fatigue, contact fatigue, and wear resistance. Studies have revealed that the three-layer sprayed graphene/DLC coatings demonstrated excellent tribological properties under vacuum conditions, achieving a friction coefficient of approximately 0.08 and a wear rate of 4.59×10^{-7} $mm^3/N{\cdot}m$. Synthetic ester-based lubricants were effective in reducing friction under extreme boundary conditions. Coatings containing organic friction modifiers maintained low friction at high temperatures, while ZnDTP provided excellent anti-wear protection across a broad temperature range, albeit with an increase in friction coefficient. Silicon-containing DLC coatings exhibited varied friction and wear characteristics, with DLC/Si (10%) achieving the lowest friction coefficient and a wear rate of 0.6×10^{-3} $mm^3/N{\cdot}m$. Although PFPE lubricants reduced friction, they increased contamination at high temperatures. The wear coefficient of DLC/biodegradable lubricants was approximately 0.1×10^{-18} m^2/N, and DLC/Ti coatings exhibited lower wear compared to pure-DLC coatings. In future development, by incorporating a variety of lubricants into the DLC coating and combining the advantages of different materials, a composite coating with excellent tribological properties and multifunctionality will be developed. Additionally, a DLC coating with self-healing ability is being developed, allowing it to self-repair through the migration and reconstruction of the internal lubricant when damaged, thereby extending the service life and reliability of the coating.

4. Conclusions and Perspectives

This paper provides a comprehensive review of five common types of self-lubricating coatings, namely MoN coatings, VN coatings, WN coatings, TMN soft-metal coatings, and DLC coatings with lubricant agents. These coatings are deposited using the PVD technique to meet the rigorous demands of tribological applications.

MoN coatings exhibit high hardness (10.6–31.4 GPa) and excellent wear resistance, particularly due to the formation of a self-lubricating oxide layer at high temperatures. VN coatings also demonstrate excellent self-lubricating properties at high temperatures, with a hardness range of 11–37 GPa. WN coatings offer good wear resistance and hardness (23–38.7 GPa), maintaining a low friction coefficient at elevated temperatures.

TMN-SMe coatings, which incorporate soft metals such as Ag, Cu, and Au, significantly enhance the self-lubricity and wear resistance of the coating. These coatings leverage the inherent lubricating properties of the soft metals to reduce friction and improve overall performance under demanding conditions.

Lubricant-doped DLC (diamond-like carbon) coatings are renowned for their low friction coefficients and excellent wear resistance. For example, W/DLC, Ni/DLC, and Cr-DLC coatings have friction coefficients of 0.2, 0.17, and 0.17, respectively. These coatings combine the hard, wear-resistant nature of DLC with the lubricating properties of the added elements, resulting in a synergistic effect that enhances their tribological performance.

Overall, each of these coatings exhibits specific advantages in different applications, particularly under high-temperature and high-load conditions. The high hardness and wear resistance of MoN, VN, and WN coatings make them suitable for demanding environments where durability is critical. TMN-SMe and lubricant-doped DLC coatings offer exceptional self-lubricating properties, making them ideal for applications where reducing friction is essential. The development and optimization of these coatings continue to advance, with research focusing on fine-tuning their compositions and structures to meet the evolving demands of various industries. By leveraging the unique properties of each coating type, it is possible to enhance the performance and longevity of mechanical components, reduce maintenance needs, and improve overall efficiency in a wide range of applications.

Designing coatings presents several challenges, including (i) achieving a balance between green technology and optimizing exceptional tribological properties; (ii) maintaining specific properties under demanding tribological conditions; and (iii) controlling relevant parameters to design improved green self-lubricating coatings.

Different views can be considered, with respect to the future development of this type of coatings. To begin with, green technology will be widely used in surface engineering; secondly, the tribological properties of the self-lubricant coatings in the field of cutting tools will be optimized and the lifespan of the equipment prolonged; finally, it is essential to analyze the feasibility of application in real industry.

However, although this type of coating has demonstrated excellent tribological properties, further research is still needed to improve its structure, and enhance environmental adaptability and functional diversity to meet the needs of a wider range of applications. This includes improved performance under different operating conditions, such as contact pressure, temperature changes, etc., as well as its reduced sensitivity to environmental changes. In addition, an in-depth understanding of the wear mechanism of coatings and their long-term stability in practical applications is also a key direction for future research.

Funding: The research was funded by National Natural Science Foundation of China with the number of 52171071 and 51801081, national funds through FCT of Portugal—Fundação para a Ciência e a Tecnologia, under a scientific contract of 2021.04115.CEECIND, 2023.06224.CEECIND, and the projects of UIDB/00285/2020, and LA/0112/2020, Post-graduate Research Innovative Training Program Jiangsu Province of China grant number SJCX23_2190.

Institutional Review Board Statement: Not applicable.

Informed Consent Statement: Not applicable.

Data Availability Statement: Data are contained within the article.

Conflicts of Interest: The authors declare no conflict of interest.

References

1. Kroll, C.; Warchold, A.; Pradhan, P. Sustainable development goals (sdgs): Are we successful in turning trade-offs into synergies? *Palgrave Commun.* **2019**, *5*, 140. [CrossRef]
2. Swain, B.; Yang-Wallentin, R. Achieving sustainable development goals: Predicaments and strategies. *Int. J. Sustain. Dev. World Ecol.* **2020**, *27*, 96–106. [CrossRef]
3. Buch, V.R.; Chawla, A.K.; Rawal, S.K. Review on electrochromic property for WO_3 thin films using different deposition techniques. *Mater. Today Proc.* **2016**, *3*, 1429–1437. [CrossRef]
4. Holmberg, K.; Erdemir, A. Influence of tribology on global energy consumption, costs and emissions. *Friction* **2017**, *5*, 263–284. [CrossRef]
5. Park, H.J.; Kim, Y.S.; Lee, Y.H.; Hong, S.H.; Kim, K.S.; Park, Y.K. Design of nano-scale multilayered nitride hard coatings deposited by arc ion plating process: Microstructural and mechanical characterization. *J. Mater. Res. Technol.* **2021**, *15*, 572–581. [CrossRef]
6. Li, W.; Liu, P.; Liaw, P.K. Microstructures and properties of high entropy alloy films and coatings: A review. *Mater. Res. Lett.* **2018**, *6*, 199–229. [CrossRef]
7. Nasim, M.; Li, Y.; Wen, M.; Wen, C. A review of high-strength nanolaminates and evaluation of their properties. *J. Mater. Sci. Technol.* **2020**, *50*, 215–244. [CrossRef]
8. Saenz-Trevizo, A.; Hodge, A.M. Nanomaterials by design: A review of nanoscale metallic multilayers. *Nanotechnology* **2020**, *31*, 292002. [CrossRef]

9. Voevodin, A.A.; Zabinski, J.S. Superhard, functionally gradient, nanolayered and nanocomposite diamond-like carbon coatings for wear protection. *Diam. Relat. Mater.* **1998**, *7*, 463–467. [CrossRef]
10. Gu, L.; Ke, P.; Zou, Y.; Li, X.; Wang, A. Amorphous self-lubricant MoS_2-C sputtered coating with high hardness. *Appl. Surf. Sci.* **2015**, *331*, 66–71. [CrossRef]
11. Vuchkov, T.; Evaristo, M.; Yaqub, T.B.; Polcar, T.; Cavaleiro, A. Synthesis, microstructure and mechanical properties of W-S-C self-lubricant thin films deposited by magnetron sputtering. *Tribol. Int.* **2020**, *150*, 106363. [CrossRef]
12. Pimentel, J.V.; Danek, M.; Polcar, T.; Cavaleiro, A. Effect of rough surface patterning on the tribology of W-S-C-Cr self-lubricant coatings. *Tribol. Int.* **2014**, *69*, 77–83. [CrossRef]
13. Mutafov, P.; Evaristo, M.; Cavaleiro, A.; Polcar, T. Structure, mechanical and tribological properties of self-lubricant W-S-N coatings. *Surf. Coat. Technol.* **2015**, *261*, 7–14. [CrossRef]
14. Ju, H.; Yao, L.; Luan, J.; Geng, Y.; Xu, J.; Yu, L.; Yang, J.; Fernandes, F. Enhancement on the hardness and oxidation resistance property of TiN/Ag composite films for high temperature applications by addition of Si. *Vacuum* **2023**, *209*, 111752. [CrossRef]
15. Kumar, C.S.; Urbikain, G.; Lacalle, L.; Gangopadhyay, S.; Fernandes, F. Investigating the effect of novel self-lubricant TiSiVN films on topography, diffusion and oxidation phenomenon at the chip-tool interface during dry machining of Ti-6Al-4V alloy. *Tribol. Int.* **2023**, *186*, 108604. [CrossRef]
16. Zhu, X.; Ma, G.; Ding, Z.; Mu, H.; Piao, Z.; Liu, M.; Guo, W.; Xing, Z.; Wang, H. Tribological properties of the WC-10Co-4Cr-4CaF_2 wear-resistant self-lubricating coating at different temperatures. *Surf. Coat. Technol.* **2023**, *475*, 130129. [CrossRef]
17. Voevodin, A.A.; O'Neill, J.P.; Zabinski, J.S. Nanocomposite tribological coatings for aerospace applications. *Surf. Coat. Technol.* **1999**, *116–119*, 36–45. [CrossRef]
18. Voevodin, A.A.; O'Neill, J.P.; Zabinski, J.S. Tribological performance and tribochemistry of nanocrystalline WC/amorphous diamond-like carbon composites. *Thin Solid Film.* **1999**, *342*, 194. [CrossRef]
19. Voevodin, A.A.; Zabinski, J.S. Supertough wear-resistant coatings with 'chameleon' surface adaptation. *Thin Solid Film.* **2000**, *370*, 223. [CrossRef]
20. Baker, C.C.; Chromik, R.R.; Wahl, K.J.; Hu, J.J.; Voevodin, A.A. Preparation of chameleon coatings for space and ambient environments. *Thin Solid Film.* **2007**, *515*, 6737–6743. [CrossRef]
21. Voevodin, A.A.; Muratore, C.; Aouadi, S.M. Hard coatings with high temperature adaptive lubrication and contact thermal management: Review. *Surf. Coat. Technol.* **2014**, *257*, 247–265. [CrossRef]
22. Baker, C.C.; Hu, J.J.; Voevodin, A.A. Preparation of Al_2O_3/DLC/Au/MoS_2 chameleon coatings for space and ambient environments. *Surf. Coat. Technol.* **2006**, *201*, 4224–4229. [CrossRef]
23. Gautam, Y.K.; Sharma, K.; Tyagi, S.; Kumar, A.; Singh, B.P. Chapter 6—Applications of green nanomaterials in coatings. Green Nanomaterials for Industrial Applications. In *Micro and Nano Technologies*; Elsevier: Amsterdam, The Netherlands, 2022; pp. 107–152.
24. Buch, V.; Oza, A.D. Machining parameter comparison on HSS cutting tool coated with multi-walled carbon nano tube using green coating technique. *Mater. Today* **2022**, *62*, 7270–7274. [CrossRef]
25. Ju, H.; Huang, K.; Luan, J.; Geng, Y.; Yang, J.; Xu, J. Evaluation under temperature cycling of the tribological properties of Ag-SiN_x films for green tribological applications. *Ceram. Int.* **2023**, *49*, 30115–30124. [CrossRef]
26. Jian, S.; Liu, Y.; Chang, C. Novel green and ecofriendly route for fabricating a robust corrosion protection coating on AZ91D magnesium alloy. *Electrochem. Commun.* **2024**, *160*, 107667. [CrossRef]
27. Fan, W.; Ding, Y.; Xiao, Z. A brand new green coating technology for realizing the regulation of spherical propellant energy release process. *Def. Technol.* **2024**, *36*, 78–94. [CrossRef]
28. Zhang, Z.; Dong, H. A State-of-the-art overview: Recent development in low friction and wear-resistant coatings and surfaces for high-temperature forming tools. *Manuf. Rev.* **2014**, *1*, 24. [CrossRef]
29. Roostaei, M.; Tabrizi, A.T.; Aghajani, H. Predicting the wear rate of Alumina/MoS_2 nanocomposite coatings by development of a relation between Lancaster coefficient & surface parameters. *Tribol. Int.* **2024**, *191*, 109129.
30. Fernandes, F.; AL-Rjoub, A.; Cavaleiro, D.; Polcar, T.; Cavaleiro, A. Room and high temperature tribological performance of multilayered TiSiN/TiN and TiSiN/TiN(Ag) coatings deposited by sputtering. *Coatings* **2020**, *10*, 1191. [CrossRef]
31. Du, J.; Lu, M.; Fang, J.; Li, W.; Chen, D. Current-carrying friction behavior and wear mechanism of Ag coatings by rotary spray deposition. *Wear* **2024**, *546–547*, 205367. [CrossRef]
32. Vitu, T.; Escudeiro, A.; Polcar, T.; Cavaleiro, A. Sliding properties of Zr-DLC coatings: The effect of tribolayer formation. *Surf. Coat. Technol.* **2014**, *258*, 734–745. [CrossRef]
33. Yaqub, T.B.; Vuchkov, T.; Bruyere, S.; Pierson, J.; Cavaleiro, A. A revised interpretation of the mechanism governing low friction tribolayer formation in alloyed-TMD self-lubricating coatings. *Appl. Surf. Sci.* **2023**, *571*, 151302. [CrossRef]
34. Selvakumar, N.; Barshilia, H.C. Review of physical vapor deposited (PVD) spectrally selective coatings for mid-and high-temperature solar thermal applications. *Sol. Energy Mater. Sol. Cells* **2012**, *98*, 1–23. [CrossRef]
35. Reichelt, K.; Jiang, X. The preparation of thin films by physical vapour deposition methods. *Thin Solid Film.* **1990**, *191*, 91–126. [CrossRef]
36. Hamzan, N.B.; Ng, C.Y.B.; Sadri, R.; Lee, M.K.; Chang, L.-J.; Tripathi, M.; Dalton, A.; Goh, B.T. Controlled physical properties and growth mechanism of manganese silicide nanorods. *J. Alloys Compd.* **2021**, *851*, 156693. [CrossRef]

37. Gutpa, J.; Shaik, H.; Kumar, K.N.; Sattar, S.A. PVD techniques proffering avenues for fabrication of porous tungsten oxide (WO3) thin films: A review. *Mater. Sci. Semicond. Process.* **2022**, *143*, 106534. [CrossRef]
38. Hidalgo, H.; Thomann, A.L.; Lecas, T.; Vulliet, J.; Wittmann-Teneze, K.; Damiani, D. Optimization of DC reactive magnetron sputtering deposition process for efficient YSZ electrolyte thin film SOFC. *Fuel Cell* **2013**, *13*, 279–288. [CrossRef]
39. Wen, J.N.; Wu, R.J.; Mi, Y.; Zhang, N.; Lin, Z.; Chen, S. Effect of Ti as cosputtering target on microstructure and mechanical properties of FeCoNi(CuAl)$_{0.2}$ high-entropy alloy thin films. *Mater. Chem. Phys.* **2023**, *296*, 127214. [CrossRef]
40. Gudmundsson, J.T. Physics and technology of magnetron sputtering discharges. *Plasma Sources Sci. Technol.* **2020**, *29*, 113001. [CrossRef]
41. Gudmundsson, J.T.; Hecimovic, A. Foundations of DC plasma sources. *Plasma Sources Sci. Technol.* **2017**, *26*, 123001. [CrossRef]
42. Kozyrev, A.V.; Sochugov, N.S.; Oskomov, K.V.; Zakharov, A.N.; Odivanova, A.N. Optical studies of plasma inhomogeneities in a high-current pulsed magnetron discharge. *Plasma Phys. Rep.* **2011**, *37*, 621–627. [CrossRef]
43. Pan, Y.; Wang, J.; Lu, Z.; Wang, R.; Xu, Z. A review on the application of magnetron sputtering technologies for solid oxide fuel cell in reduction of the operating temperature. *Int. J. Hydrogen Energy* **2024**, *50*, 1179–1193. [CrossRef]
44. Uchida, M.; Nihira, N.; Mitsuo, A.; Toyoda, K.; Kubota, K.; Aizawa, T. Friction and wear properties of CrAlN and CrVN films deposited by cathodic arc ion plating method. *Surf. Coat. Technol.* **2004**, *177*, 627–630. [CrossRef]
45. Ikeda, T.; Satoh, H. Phase formation and characterization of hard coatings in the TiAlN system prepared by the cathodic arc ion plating method. *Thin Solid Film.* **1991**, *195*, 99–110. [CrossRef]
46. Tu, R.; Yang, M.; Yuan, Y.; Min, R.; Li, Q.; Yang, M.; Ji, B.; Zhang, S. Sandwich structure to enhance the mechanical and electrochemical performance of TaN/(Ta/Ti)/TiN multilayer films prepared by multi-arc ion plating. *Coatings* **2022**, *12*, 694. [CrossRef]
47. Ghasemi, H.; Farhadi, S. A comprehensive review on the tribological and mechanical properties of nitride-based PVD coatings: Recent advancements and future directions. *Surf. Coat. Technol.* **2019**, *374*, 149–170.
48. Rodriguez, R.J.; Garcia, J.A.; Medrano, A.; Rico, M.; Sanchez, R.; Martinez, R.; Labrugere, C.; Lahaye, M.; Guette, A. Tribological behaviour of hard coatings deposited by arc-evaporation PVD. *Vacuum* **2002**, *67*, 559–566. [CrossRef]
49. Torres, H.; Ripoll, M.R.; Prakash, B. Tribological behaviour of self-lubricating materials at high temperatures. *Int. Mater. Rev.* **2018**, *63*, 309–340. [CrossRef]
50. Yang, Q.; Zhao, L.; Patnaik, P.C.; Zeng, X. Wear resistant TiMoN coatings deposited by magnetron sputtering. *Wear* **2006**, *261*, 119–125. [CrossRef]
51. Pogrebnjak, A.D.; Beresnev, V.M.; Bondar, O.V.; Zaleski, K.; Coy, E.; Jurga, S.; Lisovenko, M.O.; Konarski, P.; Rebouta, L. Superhard CrN/MoN coatings with multilayer architecture. *Mater. Des.* **2018**, *153*, 47–59. [CrossRef]
52. Xiang, J.Y.; Lin, Z.X.; Renoux, E.; Wu, F.B. Microstructure evolution and indentation cracking behavior of MoN multilayer films. *Surf. Coat. Technol.* **2018**, *350*, 1020–1027. [CrossRef]
53. Jauberteau, I.; Bessaudou, A.; Mayet, R.; Cornette, J.; Jauberteau, J.L.; Carles, P.; Merle-Mejean, T. Molybenum Nitride Films: Crystal Structures, Synthesis, Mechanical, Electrical and Some Other Properties. *Coatings* **2015**, *5*, 656–687. [CrossRef]
54. Nordin, M.; Larsson, M.; Hogmark, S. Mechanical and tribological properties of multilayered PVD TiN/CrN, TiN/MoN, TiN/NbN and TiN/TaN coatings on cemented carbide. *Surf. Coat. Technol.* **1998**, *106*, 234–241. [CrossRef]
55. Ozturk, A.; Ezirmil, K.V.; Kazmanli, K.; Urgen, M.; Eryilmaz, O.L.; Erdemir, A. Comparative tribological behaviors of TiN-, CrN- and MoN-Cu nanocomposite coatings. *Tribol. Int.* **2008**, *41*, 49–59. [CrossRef]
56. Solak, N.; Ustel, F.; Urgen, M.; Aydin, S.; Cakir, A.F. Oxidation behavior of molybdenum nitride coatings. *Surf. Coat. Technol.* **2003**, *174*, 713–719. [CrossRef]
57. Sarioglu, C.; Demirler, U.; Kazmanli, M.; Urgen, M. Measurement of residual stresses by X-ray diffraction techniques in MoN and Mo_2N coatings deposited by arc PVD on high-speed steel substrate. *Surf. Coat. Technol.* **2005**, *190*, 238–243. [CrossRef]
58. Abboudi, A.; Aissani, L.; Saoudi, A.; Djebaili, H. Effect of film thickness on the structural and tribo-mechanical properties of reactive sputtered molybdenum nitride thin films. *Metall. Mater. Eng.* **2022**, *28*, 419–433. [CrossRef]
59. Lu, X.; Su, X.; Kang, J.; Zhang, X.; Miao, X.; Wang, J.; Hao, J. Effect of peak current on the microstructure, mechanical properties and tribological behavior of Mo_xN coatings deposited by high power impulse magnetron sputtering. *Tribol. Int.* **2023**, *189*, 108955. [CrossRef]
60. Liu, C.; Ju, H.; Yu, L.; Xu, J.; Geng, Y.; He, W.; Jiao, J. Tribological Properties of Mo_2N Films at Elevated Temperature. *Coatings* **2019**, *9*, 734. [CrossRef]
61. Gassner, G.; Mayrhofer, P.H.; Kutschej, K.; Mitterer, C.; Kathrein, M. Magnéli phase formation of PVD Mo-N and W-N coatings. *Surf. Coat. Technol.* **2006**, *201*, 3335–3341. [CrossRef]
62. Qian, J.; Li, S.; Pu, J.; Cai, Z.; Wang, H.; Cai, Q.; Ju, P. Effect of heat treatment on structure and properties of molybdenum nitride and molybdenum carbonitride films prepared by magnetron sputtering. *Surf. Coat. Technol.* **2019**, *374*, 725–735. [CrossRef]
63. Fan, J.; Wang, W.; Wang, H.; Pu, J. Tribological Performance and Mechanism of MoN/VN Multilayer Films with Different Modulation Periods at Different Temperature. *Tribol. Lett.* **2021**, *69*, 154. [CrossRef]
64. Zhu, X.; Yue, D.; Shang, C.; Fan, M.; Hou, B. Phase composition and tribological performance of molybdenum nitride coatings synthesized by IBAD. *Surf. Coat. Technol.* **2013**, *228*, S184–S189. [CrossRef]
65. Samra, H.A.; Staedler, T.; Aronov, I.; Xia, J.; Jia, C.; Wenclawiak, B.; Jiang, X. Deposition and characterization of nanocrytalline Mo_2N/BN composite coatings by ECR plasma assisted CVD. *Surf. Coat. Technol.* **2010**, *204*, 1919–1924. [CrossRef]

66. Gilewicz, A.; Warcholinski, B.; Murzynski, D. The properties of molybdenum nitride coatings obtained by cathodic arc evaporation. *Surf. Coat. Technol.* **2013**, *236*, 149–158. [CrossRef]
67. Hazar, H. Characterization of MoN coatings for pistons in a diesel engine. *Mater. Des.* **2010**, *31*, 624–627. [CrossRef]
68. Zhang, G.; Fan, T.; Wang, T.; Chen, H. Microstructure, mechanical and tribological behavior of MoN_x/SiN_x multilayer coatings prepared by magnetron sputtering. *Appl. Surf. Sci.* **2013**, *274*, 231–236. [CrossRef]
69. Liang, B.; Hsieu, F.; Wu, F. Modulation effect on mechanical properties of nanolayered MoN/MoSiN coatings. *Surf. Coat. Technol.* **2022**, *436*, 128278. [CrossRef]
70. Wang, T.; Zhang, J.; Li, Y.; Gao, F.; Zhang, G. Self-lubricating TiN/MoN and TiAlN/MoN nano-multilayer coatings for drilling of austenitic stainless steel. *Ceram. Int.* **2019**, *45*, 24248–24253. [CrossRef]
71. Wang, T.; Zhang, G.; Ren, S.; Jiang, B. Effect of nitrogen flow rate on structure and properties of MoN_x coatings deposited by facing target sputtering. *J. Alloys Compd.* **2017**, *701*, 1–8. [CrossRef]
72. Krysina, O.V.; Ivanov, Y.F.; Koval, N.N.; Prokopenko, N.A.; Shugurov, V.V.; Petrikova, E.A.; Tolkachev, O.S. Composition, structure and properties of Mo-N coatings formed by the method of vacuum-arc plasma-assisted deposition. *Surf. Coat. Technol.* **2021**, *416*, 127153. [CrossRef]
73. Tillmann, W.; Kokalj, D.; Stangier, D. Impact of structure on mechanical properties and oxidation behavior of magnetron sputtered cubic and hexagonal MoN_x thin films. *Appl. Surf. Sci. Adv.* **2021**, *5*, 100119. [CrossRef]
74. Tillmann, W.; Kokalj, D.; Stangier, D. Influence of the deposition parameters on the texture and mechanical properties of magnetron sputtered cubic MoN_x thin films. *Materialia* **2019**, *5*, 100186. [CrossRef]
75. Zin, V.; Miorin, E.; Deambrosis, S.M.; Montagner, F.; Fabrizio, M. Mechanical properties and tribological behaviour of Mo-N coatings deposited via high power impulse magnetron sputtering on temperature sensitive substrates. *Tribol. Int.* **2018**, *119*, 372–380. [CrossRef]
76. Sun, L.; Mao, J.; Li, Y.; Zhou, Q.; Yuan, Z.; Xiong, X.; Fang, Q.; Gong, W.; Lu, J.; Tang, L.; et al. Microstructures and mechanical properties of MoN_x isomeric multilayer films with different number of layers deposited by DC magnetron sputtering. *Materialia* **2023**, *28*, 101714. [CrossRef]
77. Wicher, B.; Chodun, R.; Nowakowska-Langier, K.; Okrasa, S.; Trzciński, M.; Król, K.; Minikayev, R.; Skowroński, Ł.; Kurpask, Ł.; Zdunek, K. Relation between modulation frequency of electric power oscillation during pulse magnetron sputtering deposition of MoN_x thin films. *Appl. Surf. Sci.* **2018**, *456*, 789–796. [CrossRef]
78. Li, J.; Xiong, D.; Wu, H.; Zhu, H.; Kong, J.; Tyagi, R. Tribological properties of MoN layer on silver-containing nickel-base alloy at high temperatures. *Wear* **2011**, *271*, 987–993. [CrossRef]
79. Münz, W.D. Large-scale manufacturing of nanoscale multilayered hard coatings deposited by cathodic arc/unbalanced magnetron sputtering. *MRS Bull.* **2003**, *28*, 173–179. [CrossRef]
80. Mayrhofer, P.H.; Mitterer, C.; Hultman, L.; Clemens, H. Microstructural design of hard coatings. *Prog. Mater. Sci.* **2006**, *51*, 1032–1114. [CrossRef]
81. Magn'eli, A. Structures of the ReO_3-type with recurrent dislocations of atoms: 'homologous series' of molybdenum and tungsten oxides. *Acta Crystallogr.* **1953**, *6*, 495–500. [CrossRef]
82. Erdemir, A. A crystal chemical approach to the formulation of self-lubricating nanocomposite coatings. *Surf. Coat. Technol.* **2005**, *200*, 1792–1796. [CrossRef]
83. Wu, C.; Huang, J.; Yu, G. Optimization of deposition procession of VN thin films using design of experiment and single-variable (nitrogen flow rate) methods. *Mater. Chem. Phys.* **2019**, *224*, 246–256. [CrossRef]
84. Huang, J.; Wei, L.; Ting, I. Evaluation of fracture toughness of VN hard coatings: Effect of preferred orientation. *Mater. Chem. Phys.* **2022**, *275*, 125253. [CrossRef]
85. Ge, F.; Zhu, P.; Meng, F.; Xue, Q.; Huang, F. Achieving very low wear rates in binary transition-metal nitrides: The case of magnetron sputtered dense and highly oriented VN coatings. *Surf. Coat. Technol.* **2014**, *248*, 81–90. [CrossRef]
86. Qiu, Y.; Zhang, S.; Li, B.; Lee, J.-W.; Zhao, D. Influence of Nitrogen Partial Pressure and Substrate Bias on the Mechanical Properties of VN Coatings. *Procedia Eng.* **2012**, *36*, 217–225. [CrossRef]
87. Cai, Z.; Pu, J.; Lu, X.; Jiang, X.; Wang, L.; Xue, Q. Improved tribological property of VN film with the design of pre-oxidized layer. *Ceram. Int.* **2019**, *45*, 6051–6057. [CrossRef]
88. Gassner, G.; Mayrhofer, P.H.; Kutschej, K.; Mitterer, C.; Kathrein, M. A new low friction concept for high temperatures: Lubricious oxide formation on sputtered VN coatings. *Tribol. Lett.* **2004**, *17*, 751–756. [CrossRef]
89. Mu, Y.; Liu, M.; Zhao, Y. Carbon doping to improve the high temperature tribological properties of VN coating. *Tribol. Int.* **2016**, *97*, 327–336. [CrossRef]
90. Zhou, Z.; Rainforth, W.M.; Lewis, D.B.; Creasy, S.; Forsyth, J.J.; Clegg, F.; Ehiasarian, A.P. Oxidation behaviour of nanoscale TiAlN/VN multilayer coatings. *Surf. Coat. Technol.* **2004**, *177–178*, 198–203. [CrossRef]
91. Zhou, Z.; Rainforth, W.M.; Tan, C.C.; Zeng, P.; Ojeda, J.J.; Romero-Gonzalez, M.E.; Hovespian, P.E. The role of the tribofilm and roll-like debris in the wear of nanoscale nitride PVD coatings. *Wear* **2007**, *263*, 1328–1334. [CrossRef]
92. Kamath, G.; Purandare, A.P.E.Y.; Hovespian, P.E. Tribological and oxidation behaviour of TiAlCN/VCN nanoscale multilayer coating deposited by the combined HIPIMS/(HIPIMS-UBM) technique. *Surf. Coat. Technol.* **2011**, *205*, 2823–2829. [CrossRef]
93. Fateh, N.; Fontalvo, G.A.; Gassner, G.; Mitterer, C. Influence of high-temperature oxide formation on the tribological behaviour of TiN and VN coatings. *Wear* **2007**, *262*, 1152–1158. [CrossRef]

94. Fallquist, M.; Olsson, M. The influence of surface defects on the mechanical and tribological properties of VN-based arc-evaporated coatings. *Wear* **2013**, *297*, 1111–1119. [CrossRef]
95. Liu, Z.X.; Li, Y.; Xie, X.H.; Qin, J.; Wang, Y. The tribo-corrosion behavior of monolayer VN and multilayer VN/C hard coatings under simulated seawater. *Ceram. Int.* **2021**, *47*, 25655–25663. [CrossRef]
96. Grigore, E.; Ruset, C.; Luculescu, C. The structure and properties of VN-VCN-VC coatings deposited by a high energy ion assisted magnetron sputtering method. *Surf. Coat. Technol.* **2011**, *205*, S29–S32. [CrossRef]
97. Park, J.; Baik, Y. Increase of hardness and oxidation resistance of VN coatings by nanoscale multilayered structurization with AlN. *Mater. Lett.* **2008**, *62*, 2528–2530. [CrossRef]
98. Cai, Q.; Li, S.; Pu, J.; Cai, Z.; Lu, X.; Cui, Q.; Wang, L. Effect of multicomponent doping on the structure and tribological properties of VN-based coatings. *J. Alloys Compd.* **2019**, *806*, 566–574. [CrossRef]
99. Storz, O.; Gasthuber, H.; Woydt, M. Tribological properties of thermal-sprayed Magneli-type coatings with different stoichiometries (Ti_nO_{2n-1}). *Surf. Coat. Technol.* **2001**, *140*, 76–81. [CrossRef]
100. Wang, Y.; Lee, J.W.; Duh, J.G. Mechanical strengthening in self-lubricating CrAlN/VN multilayer coatings for improved high-temperature tribological characteristics. *Surf. Coat. Technol.* **2016**, *303*, 12–17. [CrossRef]
101. Aouadi, S.M.; Singh, D.P.; Stone, D.S.; Polychronopoulou, K.; Nahif, F.; Rebholz, C.; Muratore, C.; Voevodin, A.A. Adaptive VN/Ag nanocomposite coatings with lubricious behavior from 25 to 1000 °C. *Acta Mater.* **2010**, *58*, 5326–5331. [CrossRef]
102. Aouadi, S.M.; Paudel, Y.; Luster, B.; Stadler, S.; Kohli, P.; Muratore, C.; Hager, C.; Voevodin, A.A. Adaptive $Mo_2N/MoS_2/Ag$ tribological nanocomposite coat ings for aerospace applications. *Tribol. Lett.* **2008**, *29*, 95–103. [CrossRef]
103. Caicedo, J.C.; Zambrano, G.; Aperador, W.; Escobar-Alarcon, L.; Camps, E. Mechanical and electrochemical characterization of vanadium nitride (VN) thin films. *Appl. Surf. Sci.* **2011**, *258*, 312–320. [CrossRef]
104. Chun, S. Properties of VN Coatings Deposited by ICP Assisted Sputtering: Effect of ICP Power. *J. Korean Ceram. Soc.* **2017**, *54*, 38–42. [CrossRef]
105. Xu, B.; Wang, Z.; Guo, P.; Shuai, J.; Ye, Y.; Wang, A.; Ke, P. Friction behavior of VN coating under different loads. *Tribology* **2020**, *5*, 656–663.
106. Kuprin, A.S.; Gilewicz, A.; Tolmachova, G.N.; Klimenko, I.O.; Kolodiy, I.V.; Vasilenko, R.L.; Warcholinski, B. Effect of Nitrogen Pressure and Substrate Bias Voltage on Structure and Mechanical Properties of Vacuum Arc Deposited VN Coatings. *Metall. Mater. Trans. A* **2023**, *54*, 4438. [CrossRef]
107. Polcar, T.; Parreira, N.M.G. Cavaleiro. Structural and tribological characterization of tungsten nitride coatings at elevated temperature. *Wear* **2008**, *265*, 19–26. [CrossRef]
108. Parreira, N.M.G.; Carvalho, N.J.M.; Vaz, F.; Cavaleiro, A. Mechanical evaluation of unbiased W-O-N coatings deposited by d.c. reactive magnetron sputtering. *Surf. Coat. Technol.* **2006**, *200*, 6511. [CrossRef]
109. Boukhris, L.; Poitevin, J.-M. Electrical resistivity, structure and composition of d.c. sputtered WNx films. *Thin Solid Films* **1997**, *310*, 222–227. [CrossRef]
110. Tiron, V.; Velicu, I.; Porosnicu, C.; Burducea, I.; Dinca, P.; Malinsky, P. Tungsten nitride coatings obtained by HiPIMS as plasma facing materials for fusion applications. *Appl. Surf. Sci.* **2017**, *416*, 878–884. [CrossRef]
111. Patra, L.; Mallick, G.; Pandey, R.; Karna, S.P. Surface stability of WN ultrathin films under O_2 and H_2O exposure: A first-principles study. *Appl. Surf. Sci.* **2022**, *588*, 152940. [CrossRef]
112. Cui, Z.; Cai, D.; Li, Y.; Li, C.; Song, Z. WN coating of TiN electrode to improve the reliability of phase change memory. *Mater. Sci. Semicond. Process.* **2022**, *138*, 106273. [CrossRef]
113. Gachon, Y.; Ienny, P.; Forner, A.; Farges, G.; Catherine, M.C.S.; Vannes, A.B. Erosion by solid particles of W/W-N multilayer coatings obtained by PVD process. *Surf. Coat. Technol.* **1999**, *113*, 140–148. [CrossRef]
114. Javdosank, D.; Musil, J.; Soukup, Z.; Haviar, S.; Cerstvy, R.; Houska, J. Tribological properties and oxidation resistance of tungsten and tungsten nitride films at temperature up to 500 °C. *Tribol. Int.* **2019**, *132*, 211–220. [CrossRef]
115. Liao, Y.; Wu, F. Microstructure evolution and mechanical properties of refractory molybdenum-tungsten nitride coatings. *Surf. Coat. Technol.* **2024**, *476*, 130154. [CrossRef]
116. Polcar, T.; Parreira, N.M.G.; Cavaleiro, A. Tribological characterization of tungsten nitride coatings deposited by reactive magnetron sputtering. *Wear* **2007**, *262*, 655. [CrossRef]
117. Lou, B.; Moirangthem, I.; Lee, J. Fabrication of tungsten nitride thin films by superimposed HiPIMS and MF system: Effects of nitrogen flow rate. *Surf. Coat. Technol.* **2020**, *393*, 125743. [CrossRef]
118. Kumara, D.D.; Kumar, N.; Kalaiselvam, S.; Dash, S.; Jayavel, R. Substrate effect on wear resistant transition metal nitride hard coatings: Microstructure and tribo-mechanical properties. *Ceram. Int.* **2015**, *41*, 9849–9861. [CrossRef]
119. Zhao, H.; Yu, L.; Ye, F. Comparison study on the oxidation behavior of WN and WCN ceramic coatings during heat treatment. *Mater. Chem. Phys.* **2018**, *206*, 144–149. [CrossRef]
120. Tsai, Y.Z.; Duh, J.G. Tribological behavior of CrN/WN multilayer coatings grown by ion-beam assisted deposition. *Surf. Coat. Technol.* **2006**, *201*, 4266–4272. [CrossRef]
121. Wang, M.X.; Zhang, J.J.; Yang, J.; Wang, L.Q.; Li, D.J. Influence of Ar/N_2 flow ratio on structure and properties of nanoscale ZrN/WN multilayered coatings. *Surf. Coat. Technol.* **2007**, *201*, 5472–5476. [CrossRef]

122. Fox-Rabinovich, G.S.; Yamamoto, K.; Veldhuis, S.C.; Kovalev, A.I.; Shuster, L.S.; Ning, L. Self-adaptive wear behavior of nano-multilayered TiAlCrN/WN coatings under severe machining conditions. *Surf. Coat. Technol.* **2006**, *201*, 1852–1860. [CrossRef]
123. Bagdasaryan, A.A.; Pshyk, A.V.; Coy, L.E.; Kempinski, M.; Pogrebnjak, A.D.; Beresnev, V.M.; Jurga, S. Structural and mechanical characterization of (TiZrNbHfTa)N/WN multilayered nitride coatings. *Mater. Lett.* **2018**, *229*, 364–367. [CrossRef]
124. Yamamoto, T.; Kawate, M.; Hasegawa, H.; Suzuki, T. Effects of nitrogen concentration on microstructures of WN_x films synthesized by cathodic arc method. *Surf. Coat. Technol.* **2005**, *193*, 372–374. [CrossRef]
125. Samano, E.C.; Clemente, A.; Díaz, J.A.; Soto, G. Mechanical properties optimization of tungsten nitride thin films grown by reactive sputtering and laser ablation. *Vacuum* **2010**, *85*, 69–77. [CrossRef]
126. Deng, Y.; Yin, S.; Hong, Y.; Wang, Y.; Hu, Y.; Zou, G.; Kuang, T.; Zhou, K. Microstructures and properties of novel nanocomposite WN_x coatings deposited by reactive magnetron sputtering. *Appl. Surf. Sci.* **2020**, *512*, 145508. [CrossRef]
127. Yang, Y.; Qu, X.; Zhang, H.; Song, M. Positioning of interstitial carbon atoms in the deformed Fe-C system. *Mater. Today Commun.* **2023**, *34*, 105377. [CrossRef]
128. Zhang, J.; Hu, C.; Chen, L.; Kong, Y.; Mayrhofer, P.H. An ab initio guided study on the shear-induced fcc-hcp transition: A case study of (Ti, Al)N/ZrN multilayers. *Scr. Mater.* **2024**, *245*, 116064. [CrossRef]
129. Sreehari, S.; George, N.S.; Jose, L.M.; Nandakumar, S.; Subramaniam, R.T.; Aravind, A. A review on 2D transition metal nitrides: Structural and morphological impacts on energy storage and photocatalytic applications. *J. Alloys Compd.* **2023**, *950*, 169888. [CrossRef]
130. Ju, H.; Yu, L.; Yu, D.; Assempah, I.; Xu, J. Microstructure, mechanical and tribological properties of TiN-Ag films deposited by reactive magnetron sputtering. *Vacuum* **2017**, *141*, 82–88. [CrossRef]
131. Ju, H.; Yu, D.; Yu, L.; Ding, N.; Xu, J.; Zhang, X.; Zhang, Y.; Yang, L.; He, X. The influence of Ag contents on the microstructure, mechanical and tribological properties of ZrN-Ag films. *Vacuum* **2018**, *148*, 54–61. [CrossRef]
132. Mulligan, C.P.; Blanchet, T.A.; Gall, D. CrN-Ag nanocomposite coatings: Tribology at room temperature and during a temperature ramp. *Surf. Coat. Technol.* **2010**, *25*, 1388–1394. [CrossRef]
133. Li, H.; Li, J.; Kong, J.; Huang, J.; Wu, Q.; Xiong, D. Achieving high toughness and wear resistance for hard TaN-Ag films actuated by Ag. *Int. J. Refract. Met. Hard Mater.* **2023**, *111*, 106076. [CrossRef]
134. Ren, P.; Zhang, S.; Qiu, J.; Yang, X.; Wang, W.; Li, Y.; Si, Y.; Wang, G.; Wen, M. Self-lubricating behavior of VN coating catalyzed by solute Ag atom under dry friction and oil lubrication. *Surf. Coat. Technol.* **2021**, *409*, 126845. [CrossRef]
135. Xu, X.; Sun, J.; Su, F.; Li, Z.; Chen, Y.; Xu, Z. Microstructure and tribological performance of adaptive MoN-Ag nanocomposite coatings with various Ag contents. *Wear* **2022**, *488–489*, 204170. [CrossRef]
136. Ju, H.; Xu, J. Microstructure and tribological properties of NbN-Ag composite films by reactive magnetron sputtering. *Appl. Surf. Sci.* **2015**, *355*, 878–883. [CrossRef]
137. Ren, Y.; Jia, J.; Cao, X.; Zhang, G.; Ding, Q. Effect of Ag content on the microstructure and tribological behaviors of NbN-Ag coatings at elevated temperatures. *Vacuum* **2022**, *204*, 111330. [CrossRef]
138. Ren, P.; Zhang, K.; He, X.; Du, S.; Yang, X.; An, T.; Wen, M.; Zheng, W. Toughness enhancement and tribochemistry of the Nb-Ag-N films actuated by solute Ag. *Acta Mater.* **2017**, *137*, 1–11. [CrossRef]
139. Zhao, Y.; Zhao, S.; Ren, L.; Denisov, V.V.; Koval, N.N.; Yang, K.; Yu, B. Effect of substrate pulse bias voltage on the microstructure and mechanical and wear-resistant properties of TiN/Cu nanocomposite films. *Rare Met. Mater. Eng.* **2018**, *47*, 3284–3288.
140. Xu, X.; Su, F.; Li, Z. Microstructure and tribological behaviors of MoN-Cu nanocomposite coatings sliding against Si_3N_4 ball under dry and oil-lubricated conditions. *Wear* **2019**, *434–435*, 202994. [CrossRef]
141. Ezirmik, K.V.; Rouhi, S. Influence of Cu additions on the mechanical and wear properties of NbN coatings. *Surf. Coat. Technol.* **2014**, *260*, 179–185. [CrossRef]
142. Ma, K.J.; Chao, C.L.; Liu, D.S.; Chen, Y.T.; Shieh, M.B. Friction and wear behaviour of TiN/Au, TiN/MoS_2 and TiN/TiCN/a-C: H coatings. *J. Mater. Process. Technol.* **2002**, *127*, 182–186. [CrossRef]
143. He, J.L.; Wang, J.; Li, W.Z.; Li, H.D. Simulation of nacre with TiN/Pt multilayers and a study of their mechanical properties. *Mater. Sci. Eng. B* **1997**, *49*, 128–134. [CrossRef]
144. Wu, J.; Wei, P.; Liu, G.; Chen, D.; Zhang, X.; Chen, T.; Liu, H. A comprehensive evaluation of DLC coating on gear bending fatigue, contact fatigue, and scuffing performance. *Wear* **2024**, *536–537*, 205177. [CrossRef]
145. Marian, M.; Zambrano, D.F.; Rothammer, B.; Waltenberger, V.; Boidi, G.; Krapt, A.; Merie, B.; Stampfl, J.; Rosenkranz, A.; Gachot, C.; et al. Combining multi-scale surface texturing and DLC coatings for improved tribological performance of 3D printed polymers. *Surf. Coat. Technol.* **2023**, *466*, 129682. [CrossRef]
146. Dai, X.; Wen, M.; Wang, J.; Cui, X.; Wang, X.; Zhang, K. The tribological performance at elevated temperatures of MoNbN-Ag coatings. *Appl. Surf. Sci.* **2020**, *509*, 145372. [CrossRef]
147. Ren, P.; Zhang, K.; Du, S.; Meng, Q.; He, X.; Wang, S.; Wen, M.; Zheng, W. Tailoring the surface chemical bond states of the NbN films by doping Ag: Achieving hard hydrophobic surface. *Appl. Surf. Sci.* **2017**, *407*, 434–439. [CrossRef]
148. Shah, R.; Pai, N.; Khandekar, R.; Aslam, R.; Wang, Q.; Yan, Z.; Rosenkranz, A. DLC coatings in biomedical applications-Review on current advantages, existing challenges, and future directions. *Surf. Coat. Technol.* **2024**, *487*, 131006. [CrossRef]
149. Luo, G.; Xu, X.; Dong, H.; Wei, Q.; Tian, F.; Qin, J.; Shen, Q. Enhance the tribological performance of soft AISI 1045 steel through a CrN/W-DLC/DLC multilayer coating. *Surf. Coat. Technol.* **2024**, *485*, 130916. [CrossRef]

150. Wu, F.; Yu, L.; Ju, H.; Asempah, I.; Xu, J. Structural, mechanical and tribological properties of NbCN-Ag nanocomposite films deposited by reactive magnetron sputtering. *Coatings* **2018**, *8*, 50. [CrossRef]
151. Liu, C.; Ju, H.; Xu, J.; Yu, L.; Zhao, Z.; Geng, Y.; Zhao, Y. Influence of copper on the compositions, microstructure and room and elevated temperature tribological properties of the molybdenum nitride film. *Surf. Coat. Technol.* **2020**, *395*, 125811. [CrossRef]
152. Liu, E.; Zhang, J.; Chen, S.; Du, S.; Du, H.; Cai, H.; Wang, L. High temperature negative wear behaviour of VN/Ag composites induced by expansive oxidation reaction. *Ceram. Int.* **2021**, *47*, 15901–15909. [CrossRef]
153. Guo, H.; Han, M.; Chen, W.; Lu, C.; Li, B.; Wang, W.; Ji, J. Microstructure and properties of VN/Ag composite films with various silver content. *Vacuum* **2017**, *137*, 97–103. [CrossRef]
154. Donnet, C.; Erdemir, A. Diamond-like carbon films: A historical overview. In *Tribology of Diamond-like Carbon Films: Fundamentals and Applications*; Springer: New York, NY, USA, 2008; pp. 1–10. [CrossRef]
155. Hauert, R. An overview on the tribological behavior of diamond-like carbon in technical and medical applications. *Tribol. Int.* **2004**, *37*, 991–1003. [CrossRef]
156. Robertson, J. Diamond-like amorphous carbon. *Mater. Sci. Eng. R Rep.* **2002**, *37*, 129–281. [CrossRef]
157. Azzi, M.; Amirault, P.; Paquette, M.; Klemberg-Sapieha, J.E.; Martinu, L. Corrosion performance and mechanical stability of 316L/DLC coating system: Role of interlayers. *Surf. Coat. Technol.* **2010**, *204*, 3986–3994. [CrossRef]
158. Choi, J.; Nakao, S.; Kim, J.; Ikeyama, M.; Kato, T. Corrosion protection of DLC coatings on magnesium alloy. *Diam. Relat. Mater.* **2007**, *16*, 1361–1364. [CrossRef]
159. Shi, J.; Ma, G.; Han, C.; Wang, H.; Li, G.; Wei, A.; Liu, Y.; Yong, Q. The tribological performance in vacuum of DLC coating treated with graphene spraying top layer. *Diam. Relat. Mater.* **2022**, *125*, 108998.
160. Podgornik, B.; Vizintin, J. Tribological reactions between oil additives and DLC coatings for automotive applications. *Surf. Coat. Technol.* **2005**, *200*, 1982–1989. [CrossRef]
161. Tasdemir, H.A.; Wakayama, M.; Tokoroyama, T.; Kousaka, H.; Umehara, N.; Mabuchi, Y.; Higuchi, T. The effect of oil temperature and additive concentration on the wear of non-hydrogenated DLC coating. *Tribol. Int.* **2014**, *77*, 65–71. [CrossRef]
162. Kuznetsova, T.; Lapitskaya, V.; Khabarava, A.; Trukhan, R.; Chizhik, S.; Torskaya, E.; Mezrin, A.; Fedorov, S.; Rogachev, A.; Warcholinski, B. Silicon addition as a way to control properties of tribofilms and friction of DLC coatings. *Appl. Surf. Sci.* **2023**, *608*, 155115. [CrossRef]
163. Song, J.; Talukder, S.; Rahman, S.M.; Jung, Y.; Yeo, C. Comparison study on surface and thermos-chemical properties of PFPE lubricants on DLC film through MD simulations. *Tribol. Int.* **2021**, *156*, 106835. [CrossRef]
164. Vercammen, K.; Acker, K.V.; Barriga, A.V.J.; Arnsek, A.; Kalin, M.; Meneve, J. Tribological behaviour of DLC coatings in combination with biodegradable lubricants. *Tribol. Int.* **2004**, *37*, 983–989. [CrossRef]
165. Barriga, J.; Kalin, M.; Acker, K.V.; Vercammen, K.; Ortega, A.; Leiaristi, L. Tribological performance of titanium doped and pure DLC coatings combined with a synthetic bio-lubricant. *Wear* **2006**, *261*, 9–14. [CrossRef]
166. Persson, K.; Gahlin, R. Tribological performance of a DLC coating in combination with water-based lubricants. *Tribol. Int.* **2003**, *36*, 851–855. [CrossRef]
167. Ueda, M.; Kadiric, A.; Spikes, H. Wear of hydrogenated DLC in MoDTC-containing oils. *Wear* **2021**, *474–475*, 203869. [CrossRef]
168. Yoshida, Y.; Kunitsugu, S. Friction wear characteristics of diamond-like carbon coatings in oils containing molybdenum dialkyldithiocarbamate additive. *Wear* **2018**, *414–415*, 118–125. [CrossRef]
169. Pei, Y.; Bui, X.; Zhou, X.; Hosson, J. Tribological behavior of W-DLC coated rubber seals. *Surf. Coat. Technol.* **2008**, *202*, 1869–1875. [CrossRef]
170. Wei, X.; Cao, X.; Yin, P.; Ding, Q.; Lu, Z.; Zhang, G. Design of a novel superhydrophobic F&Si-DLC film on the internal surface of 304SS pipes. *Diam. Relat. Mater.* **2022**, *123*, 108852.
171. Zou, C.; Wang, H.; Feng, L.; Xue, S. Effects of Cr concentrations on the microstructure, hardness, and temperature-dependent tribological properties of Cr-DLC coatings. *Appl. Surf. Sci.* **2013**, *286*, 137–141. [CrossRef]
172. Bui, X.; Pei, Y.; Hosson, J. Magnetron reactively sputtered Ti-DLC coatings on HNBR rubber. The influence of substrate bias. *Surf. Coat. Technol.* **2008**, *202*, 4939–4944. [CrossRef]
173. Pei, Y.; Bui, X.; Hosson, J. Deposition and characterization of hydrogenated diamond-like carbon thin films on rubber seals. *Thin Solid Film.* **2010**, *518*, S42–S45. [CrossRef]
174. Jing, P.; Gong, Y.; Deng, Q.; Zhang, Y.; Huang, N.; Leng, Y. The formation of the "rod-like wear debris" and tribological properties of Ag-doped diamond-like carbon films fabricated by a high-power pulsed plasma vapor deposition technique. *Vacuum* **2020**, *173*, 109125. [CrossRef]

Disclaimer/Publisher's Note: The statements, opinions and data contained in all publications are solely those of the individual author(s) and contributor(s) and not of MDPI and/or the editor(s). MDPI and/or the editor(s) disclaim responsibility for any injury to people or property resulting from any ideas, methods, instructions or products referred to in the content.

MDPI AG
Grosspeteranlage 5
4052 Basel
Switzerland
Tel.: +41 61 683 77 34

Coatings Editorial Office
E-mail: coatings@mdpi.com
www.mdpi.com/journal/coatings

Disclaimer/Publisher's Note: The title and front matter of this reprint are at the discretion of the Guest Editors. The publisher is not responsible for their content or any associated concerns. The statements, opinions and data contained in all individual articles are solely those of the individual Editors and contributors and not of MDPI. MDPI disclaims responsibility for any injury to people or property resulting from any ideas, methods, instructions or products referred to in the content.

www.ingramcontent.com/pod-product-compliance
Lightning Source LLC
Chambersburg PA
CBHW061625210926
51850CB00009B/2438
9783725847976